Biology of Fishes

Q. BONE, M.A., D.Phil.
Senior Principal Scientific Officer
Laboratory of the Marine Biological Association
Plymouth, UK

and

N. B. MARSHALL, M.A., Sc.D., F.R.S.
Emeritus Professor of Zoology
University of London

Illustrated by Q. Bone

Blackie

Glasgow and London

Distributed in the USA by
Chapman and Hall
New York

Blackie & Son Limited
Bishopbriggs, Glasgow G64 2NZ
Furnival House, 14–16 High Holborn, London WC1V 6BX

Distributed in the USA by
Chapman and Hall
in association with Methuen, Inc.,
733 Third Avenue,
New York, N.Y. 10017

British Library Cataloguing in Publication Data

Bone, Q.
    Biology of fishes. — (Tertiary level biology)
    1. Fish
    I. Title   II. Marshall, N. B.   III. Series
    597   QL617.2

    ISBN 0-216-91017-X

Library of Congress Cataloging in Publication Data

Bone, Q.
    Biology of fishes.

    (Tertiary level biology)
    Includes bibliographical references and
index.
    1. Fishes.   2. Fishes—Physiology.
I. Marshall, Norman Bertram.   II. Title.
III. Series.
QL615.B674   1982   597   82-7919
ISBN 0-412-00151-9   AACR2

Filmset by Advanced Filmsetters (Glasgow) Ltd

Printed in Great Britain by
Thomson Litho Ltd, East Kilbride, Scotland

Preface

Fishes are the most abundant and yet least known class of vertebrates. Around 20 000 species are recognized at present, and it is likely that the final total will show three in every five vertebrates to be a fish. They inhabit every kind of aquatic environment, and their wide distribution has resulted in many different designs for their special modes of life.

This book discusses the different kinds of fishes, how they operate and where they live. It is written especially for advanced undergraduates and postgraduates, but it is hoped that it will appeal to a much wider readership. In a compass of 270 pages, it has been necessary to be selective, but we have chosen those topics which seem to us to be most useful and interesting. We hope that it will introduce the reader to the fascination of this class of vertebrates and induce him to investigate more closely the complexities of what are, after all, rather little-known animals.

Q.B.
N.B.M.

This book is for Susan and Olga

Contents

CHAPTER ONE

THE ABUNDANCE AND DIVERSITY OF FISHES

1.1 Introduction

Fishes are the most abundant and yet the least known class of vertebrates.
They inhabit every kind of aquatic environment and some even spend
most of their time out of water. Their size, shape, and internal structure are
diverse. At present, about 20000 species (almost half of all known
vertebrates) are recognized, and each year about 100 new species are
described. The final total will probably be around 30000 species, which
will mean that three in every five vertebrates is a fish. The majority of these
are bony fishes, mainly teleosts, and the remainder comprises some 50
species of jawless fishes and about 800 of cartilaginous fishes.

This huge array of species is irregularly distributed—no less than four
out of ten fish species live in fresh water, although fresh waters make up
only 0.0093% of the total water on earth (Figure 1.1). Cohen (1970) has
estimated that 33% of all fishes belong to primary freshwater species,
mainly carps, characins and catfishes. Evidently the chances of isolation
and formation of new species are much greater in freshwater habitats than
in the sea. Half the marine species known are from the warm waters
fringing the land, particularly where corals are most productive, but fishes
have colonized all marine environments and have developed remarkable
adaptations to enable them to make a living at the bottom of the deepest
oceans as well as around the shores.

This wide distribution has resulted in many different designs for special
modes of life and we can study the different solutions to the same problems
that have been devised by diverse species by considering the many different
kinds of fishes alive today. For example, we can compare ways of acquiring
sufficient oxygen in lampreys and elasmobranchs, or the relative merits of
fat and gas as sources of static lift in elasmobranchs and teleosts. But

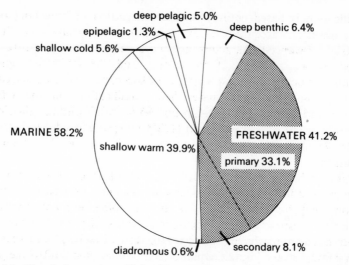

Figure 1.1   The relative numbers of living fishes in different habitats. After Cohen (1970).

before making these often illuminating comparisons, we should have a clear idea of the diversity of fishes, their basic structural features, and how they are related to each other.

Fortunately, all fishes seem to have been derived from a common ancestor and are monophyletic. Such specialized features of the nervous system as the paired giant Mauthner cells (Chapter 10) or the pattern of large associative neurones found in the spinal cord of all groups convince us of a common descent. However, it is not easy to summarize the classification of fishes, because various conflicting schemes have recently been proposed, linking the main fish groups in different ways. The aim of the next section is to give some idea of the reasons underlying these different arrangements.

1.2 The classification of fishes

Taxonomists can themselves be classified according to their method of approach; the two most influential schools at present are the cladist and the evolutionist. Cladists see evolution as a sequence of dichotomies, where a parent species gives rise to two daughter species and itself disappears, the two daughter species and all their descendants being termed 'sister groups' and having equivalent rank in the classification. Gradual modification along a single line has no place in cladistic taxonomic schemes;

classification is based entirely on the recognition of branching points where novel characteristics appear and new sister groups arise. In this approach, advanced characters are more helpful than primitive ones shared by all members of the group, and the latter are ignored. Evolutionist taxonomists, on the other hand, consider both advanced and primitive characters, and base their classification on the inferred evolutionary history of the group. This brief outline should be supplemented by reference to Mayr (1981) who provides a fuller discussion of the merits of each approach and a suggestion that they should be synthesized.

Many, perhaps most, modern fish systematists are cladists, and this has profoundly influenced fish classifications. The chordate classification of Nelson (1969), for instance, differs from that in older texts and raises many interesting questions (Figure 1.2). As in previous schemes, the isolation of jawless fishes from all other vertebrates is recognized by placing them in a separate superclass (Agnatha) which forms a sister group to the jawed

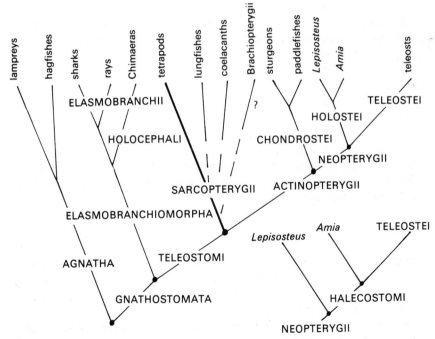

Figure 1.2 Relationships of living fish groups, after Nelson (1969). Lower right: another arrangement of holosteans and teleosts, after Patterson (1973).

vertebrates (Gnathostomata). The gnathostomes themselves are divided into two more sister groups, forming classes of equal rank. One, the Elasmobranchiomorpha, contains the cartilaginous fishes (Holocephali and elasmobranchs); the other (Teleostomi) contains all other vertebrates, including the tetrapods. This arrangement reflects the view that bony fishes and tetrapods are more closely related than are bony fishes and cartilaginous fishes. ('Teleostome' simply means 'completed jaw' and refers to the dermal bones composing the jaws of bony fishes and tetrapods, which differ from the elasmobranch jaw with its palatoquadrate and Meckel's cartilages.)

Within the Teleostomi, Nelson separates the ray-finned Actinopterygii from lungfishes, the coelacanth, and the bichir *Polypterus* and its relatives, which are all grouped with the tetrapods in the Sarcopterygii. The query in Figure 1.2 indicates that the position of the brachiopterygians (*Polypterus*) is not yet clear. For simplicity we shall class them with the Chondrostei as actinopterygians of a relatively primitive grade of organization, although some current opinion (see Jarvik, 1980) is that they are more closely related to the crossopterygians.

Figure 1.2 shows one way in which fishes can be grouped to show their assumed phyletic relationships. However, the expression of such relationships is only one possible function of a classification, and it may sometimes be more useful to apply an 'artificial' classification. For example, gnathostome fishes may be divided from the tetrapods by placing them in a grade Pisces, including Elasmobranchiomorpha and Osteichthyes as two classes, equivalent to the four classes of tetrapods (Nelson, 1976). This is the arrangement found in older texts, where the Osteichthyes are often separated into two subclasses: Actinopterygii (ray-fins) and Sarcopterygii (lobe-fins—the lungfishes and coelacanth, but not *Polypterus*) to emphasize the differences between these two kinds of bony fish. The Actinopterygii in turn were originally subdivided on the evidence of their living representatives, and so three quite distinct ray-finned fish groups were proposed: Chondrostei (based on sturgeons), Holostei (*Amia* and *Lepisosteus*) and Teleostei. However, when these groups were traced back and expanded to include their fossil relatives, it became clear that their original significance and phylogenetic validity was in question. Teleosts still seem to be monophyletic, but the other two groups represent successive levels of organization rather than single lines of descent (Moy-Thomas and Miles, 1971). So the Chondrostei may be regarded as actinopterygian fishes at the earliest level of organization, and the Holostei as an intermediate grade, some of which gave rise to the great adaptive

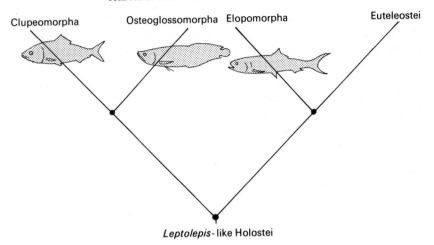

Figure 1.3 Relationships of major teleost groups, after Greenwood (1973). Note that Clupeomorpha and Osteoglossomorpha together form the sister group to all other teleosts.

radiation of the Teleostei which dominated the fish fauna from the time of the Upper Cretaceous.

These considerations have led several recent fish taxonomists to adopt a classification of actinopterygians which recognizes that at present it is more convenient to group them in organizational levels rather than on cladistic principles. Differences between existing fish classifications are thus not only based on different taxonomic approaches but also on different ideas about the most convenient way to arrange a working classification.

Within the large groups shown in Figure 1.2, attempts to arrange the subgroups to display their relationships are still only provisional, so here too older and more recent texts are apt to differ. Perhaps the greatest difficulties arise in dealing with the vast assemblage of teleosts. Greenwood (1973) has linked the freshwater osteoglossomorph fishes (osteoglossids and mormyrids) with the Clupeomorpha (herrings and anchovies) and suggests that the two together form one sister group to all other teleosts (Figure 1.3). The other teleosts were divided into Elopomorpha (tarpons, eels and other fishes with leptocephali larvae) and the great radiation of the Euteleostei. The modern classification of the euteleosts themselves, outlined in Figure 1.4, began with the influential paper of Greenwood and his colleagues (1966).

All familiar teleosts, except herrings and eels, belong to the five euteleost superorders. The first, Protacanthopterygii, contains salmon, trout, and a few deep-sea forms. The second, Ostariophysi, is much larger and includes characins, cyprinids and catfishes. The third superorder, Scopelomorpha, contains mainly deep-sea fishes, including the large radiation of

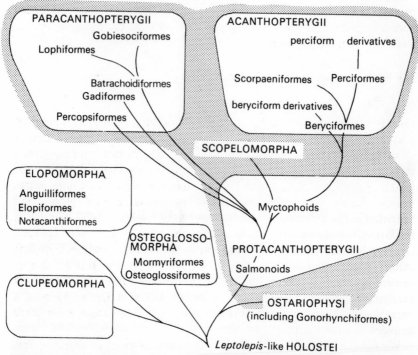

Figure 1.4 Relationships of euteleost groups (within shaded outline) and other living teleosts. Modified from Greenwood *et al.* (1966).

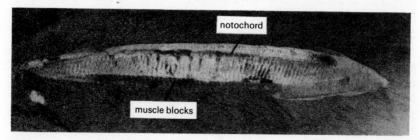

Figure 1.5 The Lower Cambrian *Pikaia* from the Burgess shales (photograph kindly provided by Dr S. Conway Morris).

lanternfishes (myctophids) many of which have curiously modified eyes and luminous organs. Fourth, distinguished by a particular kind of jaw musculature and caudal skeleton, are the Paracanthopterygii, including gadoids (cod and haddock), anglerfishes, and a variety of other forms. Finally, much the most important euteleost radiation is the Acanthopterygii, recognized by the spiny rays in the fins. This most advanced superorder includes the great majority of teleosts, for example wrasse, swordfish, tunas and perches.

1.3 Basic structural features of fishes

1.3.1 Body shape and fins

During the evolution of fishes, many wide-ranging variations have been played on ancestral themes of form and organization. As a result, fishes are so varied that they are hard to define. Ancient fish ancestors were typically spindle-shaped and streamlined, as we see from a fortunate glimpse of the Middle Cambrian fauna in the Burgess shales of Western Canada, where *Pikaia* (Figure 1.5) shows the zigzag myotomal muscle blocks and notochord characteristic of all fishes. Many modern fishes retain the fusiform streamlined body, but in addition every other kind of body shape is represented (Figure 2.7): near-globular (pufferfishes and some anglerfishes), elongate (eels, hagfish and lampreys), or compressed dorsoventrally (rays) or laterally (sunfishes and many coral-reef fishes). This extraordinarily wide range in body shape can best be appreciated around a coral reef, but similar diversity is found in the deep ocean or in large tropical rivers.

Whatever their body shape, most fishes have a series of unpaired fins, together with paired pectoral and pelvic fins. Modern agnathans lack the paired fins, although some fossil agnathans had them, and it seems probable that those of gnathostomes arose from lateral dermal folds, seen on certain Upper Silurian anaspids (Moy-Thomas and Miles, 1971). The development of paired fins conferred the ability for delicate and rapid manoeuvre, and fish fins show a great diversity of form and structure according to the mode of life. In sharks, the fins are stiffened at the base by cartilaginous plates to which is attached a series of smaller radial cartilages (Figure 1.6). From the radial cartilages, bristle-like collagenous rays (ceratotrichia) pass outwards to the edge of the fin. This kind of fin cannot be furled, but muscles attached to the basal cartilages can tilt the fin. In rays, the radials of the pectorals and pelvics are much elongated and jointed, so that muscles inserting on them flex the pectorals up and down

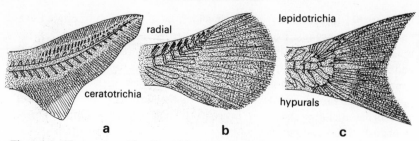

Figure 1.6   Fin structure. After Goodrich (1909). (*a*) Shark; (*b*) *Lepisosteus*; (*c*) *Salmo*.

as the ray wings its way along. But it is in teleosts that the most flexible and versatile fin musculature is found. Here, flexible jointed rays (lepidotrichia) are supplied with basal muscles which enable the fish to fold the fins (Figure 1.6) and to make delicate propulsive movements whose precision is astonishing (perhaps the most striking examples are seen in the seahorses and knifefishes). Sometimes, as in sea robins and gurnards, free pectoral fin rays are used for sampling the bottom as the fish walks forwards. However, not all teleost fishes have flexible fins. For example, in the pectoral fins of marlins the fin rays are ossified together to form a very stiff lifting foil, rather similar to that of large sharks, except that it can be folded back against the body.

1.3.2 Internal features

The bodies of all fishes are built around an axial notochord (more or less reduced in most by the development of vertebral elements) which is flexed by the serially arranged myotomal muscle blocks. All have gills for gas exchange, a heart and circulatory system of essentially the same plan, and a similar complement of viscera. However, in internal structure agnathan fishes are very different to gnathostome fishes (Figures 1.7–1.9). For example, the branchial skeleton of lampreys and hagfish is attached to the cranium and lies outside the gill pouches, whilst in other fishes the branchial skeleton is independent and supports the gill arches directly. Living agnathans have only a single nasal opening leading to a nasohypophyseal tract, unlike the gnathostome paired olfactory organs. Still other differences are the lack of paired fins, of bone or dentine, of a lymphatic system, and of myelin sheaths around the nerve fibres of agnathans.

   Much the most important difference between the two groups is that gnathostomes have hinged jaws, and it is certainly this development which accounts for their overwhelming success compared with the very few

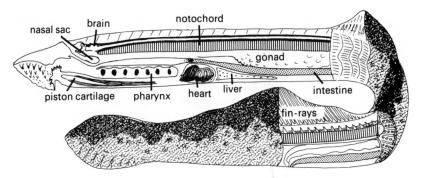

Figure 1.7   General features of a lamprey. After Goodrich (1909).

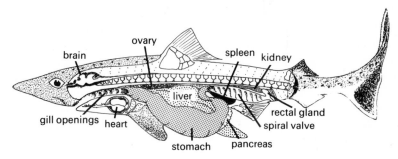

Figure 1.8   General features of a shark (*Squalus*). After Lagler *et al.* (1977).

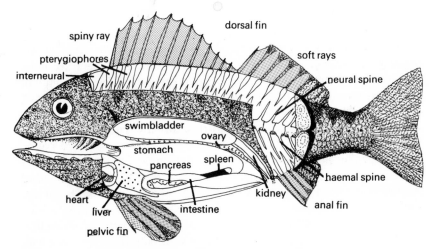

Figure 1.9   General features of a teleost (*Perca*). After Dean (1895).

modern survivors of the agnathan radiations between the Ordovician and Devonian periods. Together with the development of jaws went that of fusiform bodies with appropriate patterns of fins for accurate control of the movements needed to snap up larger and faster prey, and the development of stronger skeletal elements to withstand the forces of the more elaborate and stronger muscles. The gnathostomes have larger brains than the living agnathans (Chapter 10) particularly the cerebellum, which is small in agnathans or even (in hagfishes) apparently lacking. Bony fishes have a swimbladder whose development allows much more precise and efficient locomotion. Once jaws had developed, then, a whole series of changes from the agnathan level of organization became possible, leading to the dominance of the new gnathostome fishes.

Bony fishes and elasmobranchiomorphs also differ in internal structure (Figures 1.8, 1.9). In elasmobranchiomorphs, the skull in the adult remains as a cartilaginous neurocranium to which are fused cartilaginous capsules surrounding the olfactory organs and the ear; to this bulbous and sometimes bizarrely-shaped braincase are attached the palatoquadrate and Meckel's cartilages of the upper and lower jaw (see Figure 7.5). The vertebral column is built from circular centra, slightly concave on either face and separated by intervertebral discs; dorsally, the centra bear neural arches, and ventrally either short ribs or, in the tail region, haemal arches. This rather simple axial skeleton contrasts with the more complex skull and vertebral column of bony fishes.

In the bony fishes, the skull begins development as a cartilaginous neurocranium, but it is altered by ossifications in the cartilage, forming endochondral bones, and by the addition of many membrane or dermal bones, supposedly derived from a scale layer in the skin. The skull of bony fishes is therefore a dauntingly complex structure, and there are great difficulties in homologizing dermal bones between different groups of fishes. For instance, the large paired dermal bones above the teleost orbit, under which the pineal lies, were for a long time labelled frontals (as they still are in some recent texts) whereas they are now known to be equivalent to the tetrapod parietals.

Space precludes a detailed description of the bony fish skull (see Lagler et al., 1977, or Young, 1981). Here, the relative roles of the endochondral and dermal bones are illustrated by examining the jaws. As in the elasmobranchiomorphs, the jaws of bony fishes begin as palatoquadrate and Meckel's cartilage. Centres of ossification in the palatoquadrate give rise to palatine, pterygoid, mesopterygoid and quadrate bones (some with dermal contributions), the edges of the upper jaw being formed from membrane bones (premaxilla, maxilla and jugal) which cover the endo-chondral elements. In the lower jaw, the only ossification in Meckel's cartilage is at its hinder end, forming the articular; the dermal dentary forms the major part of the jaw.

The vertebral column in bony fishes is usually similar to the elasmo-branchiomorph design, with biconcave centra but more complex ribs. However, in some groups, such as sturgeons, and in *Latimeria*, the notochord remains as a massive unconstricted rod in the adult.

Elasmobranchiomorphs lack a swimbladder, and the reproductive system is different to that in bony fishes (fertilization is always internal). In other respects, the internal organs in both groups are similar. Bony fishes often have gut diverticula, which are lacking in elasmobranchiomorphs. A few bony fishes have spiral valves (e.g. *Amia*, *Polypterus* and osteoglossids) but these are less complex than those of elasmobranchiomorphs, and with the exception of *Latimeria*, bony fishes lack the salt-secreting rectal glands characteristic of elasmobranchiomorphs.

1.4 Distribution and morphology

After this brief outline, we may now consider the different groups of fishes in more detail.

1.4.1 Agnatha

Lampreys and hagfish look superficially alike, but are not closely related. The existence of an undoubted fossil lamprey (*Mayomyzon*, of the Upper Carboniferous) implies that their ancestors lived in the Palaeozoic (Bardack and Zangerl, 1971). Today, lampreys are found in rivers north and south of latitude 30° N and S; all but four species occur only in the Northern Hemisphere. Most species are confined to fresh water, where they spawn and their ammocoete larvae burrow in the river beds, but some species (e.g. *Geotria australis* and *Petromyzon marinus*) migrate to the sea after metamorphosis. These anadromous forms, and also some of the freshwater species, feed on other fishes by rasping the skin with a tooth-studded muscular tongue, but an equal number of species do not feed at all as adults and are smaller than the corresponding parasitic species. Such 'dwarf' forms probably evolved from parasitic species, perhaps as a result of isolation in reaches of river systems where host fishes were uncommon, or where for other reasons mortality was high (Hardisty, 1979).

The 20 or so species of hagfish are entirely marine, living in the sea bed, mostly in antitropical regions down to slope levels of 2000 m or more. Unlike lampreys, development is direct, the advanced young emerging from large yolky eggs. Hagfish feed on small invertebrates such as polychaete worms, and on dead fishes, exuding large quantities of slime from serial slime glands, which deters others from joining in the feast. Their remarkably flexible bodies, which they can tie into knots, are unique in this ability.

Compared with lampreys, hagfishes have a reduced sensory system. For example, the small eyes lie under pigmented skin, and there is only a single semi-circular canal in the ear (there are two in lampreys). The lateral line system is poorly developed, and electroreceptors are unknown.

1.4.2 Elasmobranchiomorpha

The thirty or so species of chimaeroid fishes (Holocephali) and the 700–800 sharks and rays form a group that has long been regarded as illustrating a general or 'primitive' fish design, partly because the skeletal and visceral morphology are relatively simple and show features found in the ontogeny of other groups. But it is now clear that living elasmobranchiomorphs are not in fact 'simple'; as Moss (1977) emphasizes, the diversity of their feeding mechanisms is unparalleled by any other group with so few living representatives. Also, shark brains may be relatively large (Chapter 8). Elasmobranchiomorphs conserve urea for osmoregulation, and have internal fertilization. Both these features also occur in *Latimeria*, but are more advanced in the elasmobranchiomorphs and almost certainly had an independent origin. All species are marine, apart from a few sharks and rays that enter fresh water and some rays that are confined to fresh water (Chapter 5).

In both groups of elasmobranchiomorphs, the cartilaginous skeleton is strengthened at strategic points by granular or prismatic calcifications (Figure 1.10). These are less prominent in holocephalans, which are delicate 'watery' fishes, and unlike sharks and rays have a naked skin bearing denticles only on the claspers. The denticles which cover sharks and rays arise from a plate-like base where true bone tissue may be found (Moss, 1977). This supports the body of the denticle, which is made of dentine capped by a layer of enameloid material. As the fish grows, the denticles do not increase in size at the same rate and so new denticles are intercalated between the old ones. They are variously modified to form flattened scales, spines and filtering combs (as in the basking shark, *Cetorhinus*); more importantly, on the jaws they form teeth. Shark teeth

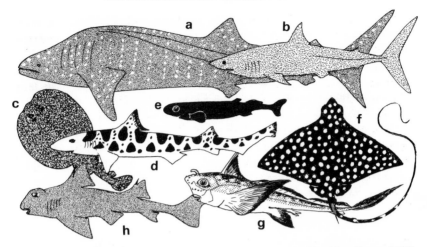

Figure 1.10 Various elasmobranchiomorphs (not to same scale). After Dean (1906), Linaweaver and Backus (1970), Daniel (1922) and Marshall (1971). (*a*) Whale shark (*Rhincodon*); (*b*) mako (*Isurus*); (*c*) electric ray (*Torpedo*); (*d*) cat shark (*Triakis*); (*e*) *Squaliolus*; (*f*) eagle ray (*Aetiobatis*); (*g*) ratfish (*Chimaera*); (*h*) *Heterodontus*.

are replaced from behind forwards, although only one row is normally in use. Dogfish use several rows at once, as do rays, where the denticles form grinding batteries. The most striking development is seen in the slashing teeth of the tiger-shark *Galeocerdo* and in similar formidable predatory sharks. In holocephalans, which have a varied diet, the jaws bear curious toothplates somewhat like rodent incisors.

Just over half of living elasmobranchiomorphs are rays (Compagno, 1977) characterized by ventral gill openings and by the fusion of the pectoral fins to the head. Most have large pectorals, like *Raja*, the main skate genus, but some, like the sawfishes (Pristidae), have much smaller ones and look much like sharks. Rays range from 10 cm or so across the wings to the 6 m of the large planktivorous myliobatids *Manta* and *Mobula*. Some bear venomous spines at the base of the tail, and others have powerful electric organs in the pectorals. All rays have weak electric organs in the tail whose function remains mysterious. Like sharks, rays also have an elaborate electroreceptor system (Chapter 7).

A quarter of living sharks are placed in the superorder Squalomorphi which, besides squaloids like the spur dog (*Squalus*) and its deep-sea relatives, also includes the pristiophorid saw-sharks and the six- and seven-gilled hexanchoid sharks. Most squaloids have relatively light organs, and the deep-sea forms have large livers used as buoyancy tanks (Chapter 4).

The smallest known sharks are squaloids, for instance *Squaliolus laticaudus*, which lives in deep waters off continental land masses and feeds on midwater squid and fishes (Seigel, 1978). *Squaliolus* is the same length as a medium-sized teleost (up to 24 cm) and it is puzzling that no sharks of this size-range are found anywhere else in the oceans. Almost all other sharks are placed in the superorder Galeomorphi, which contains some 200 species of carcharhinids such as the tiger, grey and hammerhead sharks, as well as the advanced, warm-blooded, fast-swimming isurids culminating in the great white shark *Carcharodon*. Like deep-sea squaloids, *Carcharodon* has an immense buoyant liver (the biggest white shark reliably recorded was 6.5 m long and weighed 3300 kg, the liver weighing 456 kg. A great white captured in the Azores in 1981 was no less than 9 m long, but this record awaits ratification.) The largest shark, the planktivorous *Rhincodon* (the whale-shark), grows to over 15 m, and is grouped with the carpet and nurse-sharks (orectolobids) of warm shallows of the Indopacific and Red Sea. Finally, the galeomorphs also include the large family of smooth dogfishes (triakids) like the smooth hound *Mustelus*—these are usually coastal forms found in temperate to tropical regions.

1.4.3 Actinopterygii

1. *Chondrostei*. Two small groups of chondrostean fishes survive (Figure 1.11): the 20 or so sturgeons and paddlefishes, and the eight African species of bichirs and reedfishes (*Polypterus, Calamoichthys*). Both groups show primitive characteristics such as a functional spiracle and a spiral valve in the intestine, but *Polypterus* is much closer to the original chondrosteans in appearance. It is covered with closely-set rhomboid scales, which lie within the skin and grow by addition on both surfaces, the outer surface being formed from the dense shiny substance ganoin, as in the fossil forms. Indeed, *Polypterus* seems little changed from the Devonian paleoniscoid chondrosteans (Moy-Thomas and Miles, 1971). The paired lung-like swimbladder, connected to the oesophagus by a ventral glottis, is thought to show the ancestral condition; bichirs must use their swimbladder for respiration, and die if denied access to the surface to gulp air. In sturgeons, the single swimbladder is connected with the oesophageal roof and is not used as a lung.

Polypterus is well ossified, but sturgeons have a mainly cartilaginous skeleton with an enormous unconstricted notochord. Some dermal elements in the skull, and a line of lateral scutes along the body, are ossified, and the scales of the tail region have lost the ganoin layer and consist of concentric layers of bone, as in teleosts.

Sturgeons feed mainly on bottom-living invertebrates, rooting them out with the snout and sucking them in with the toothless, protrusible ventral mouth. Some species grow very large (5 m in length and 1000 kg in weight) and then may also include other fishes like lampreys in their diet. As well as the four barbels in front of the mouth, sturgeons have ampullary electroreceptors on the snout (Chapter 7) and probably detect their prey mainly with this system. Similar electroreceptors are found on the snout of the paddlefish (*Polyodon*) of the Mississippi, and in *Psephurus* (a Chinese genus); both filter plankton from the water with a gill-raker sieve. It is curious that paddlefishes, like whale-sharks and basking sharks, are equipped with electroreceptors yet are plankton feeders. Perhaps the system is used in all three groups to detect muscle action potentials emitted by high concentrations of crustacean prey.

Some sturgeons, like the sterlet, are virtually river fishes, but most live in the oceans or in large inland seas and lakes, ascending rivers only to spawn. Fertilization is external, and the sticky eggs attach to the river bed and hatch into spiny larvae. These spend a year or more in the river after metamorphosis, and they then pass down to the sea.

2. *Holostei.* Only eight holosteans survive: the bowfin, *Amia*, of eastern North America, and the garpikes (*Lepisosteus*, four species, and *Altractosteus*, three species) from fresh and brackish waters of eastern North America and Central America. The largest species, the giant tropical alligator gar (*L. tristoechus*) grows to a length of some 3.5 m. In most characters, *Amia* and *Lepisosteus* (Figure 1.11) are much more like teleosts than are chondrosteans. The skeleton is strongly ossified and the fins are flexible, with fewer fin rays than in chondrosteans. There is no spiracle (traces remain of a spiracular pit in the pharynx near the otic capsule). Although the heart has a large conus arteriosus, and there is a reduced spiral valve in the intestine, these primitive characters are over-shadowed by such teleost features as the development of an eye-muscle canal (myodome) in the floor of the skull; the loss of the clavicle from the pectoral girdle; and the freeing of the maxillary from the preoperculum which straps together the bones supporting the jaws. This last feature allows a larger gape and protrusion of the lower jaw so that prey can be sucked in, as in many teleosts (Chapter 7). Large septate airbladders confer neutral buoyancy, as well as being used for respiration. *Amia* is covered with thin teleost-like bony scales, but *Lepisosteus* still retains the thick rhombic ganoin-plated scales, socketed together, which make the body

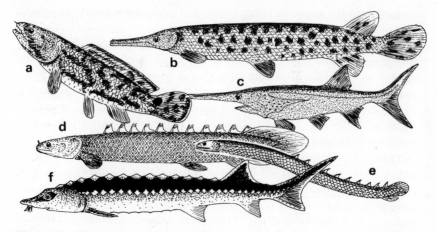

Figure 1.11 Chondrosteans and holosteans (not to same scale). After Goodrich (1909) and Marshall (1971). (*a*) Bowfin (*Amia*); (*b*) garpike (*Lepisosteus*); (*c*) paddlefish (*Polyodon*); (*d*) bichir (*Polypterus*); (*e*) reedfish (*Calamoichthys*); (*f*) sturgeon (*Acipenser*).

relatively inflexible—*Lepisosteus* catches its fish prey from ambush by a sideways snap of the jaws. In the reproductive system, however, *Amia* is more primitive than *Lepisosteus*—the ovary is not continuous with the oviduct (as in *Lepisosteus* and most teleosts) but resembles that of the elasmobranchiomorphs and chondrosteans in the way in which the eggs are shed into the body cavity before entering the oviduct. *Amia* eggs are small and undergo total cleavage, whereas in *Lepisosteus* cleavage is partial (meroblastic) as in teleosts.

3. *Teleosti.* Teleosts are immensely varied and have been the most successful fish group since they began to radiate in the Cretaceous, probably arising from halecostome holosteans similar to the leptolepids. The many 'loss' and 'gain' characters that marked the evolution of the teleosts are analysed by Patterson (1967, 1975, 1977), whose conclusion that the group is monophyletic is supported by Avise and Kitto's (1973) genetic studies.

In comparison with holosteans, teleosts are generally more active, faster-swimming, lighter fishes, though some have adopted inactive ways of life, like trunkfishes which are enclosed in a heavy bony cuirass. Some others, notably deep-sea forms, have reduced skeletal calcification which saves weight (even such an immense fish as the ocean sunfish *Mola* is very little calcified), but most teleosts have a well-calcified skeleton. The skeletal elements are much more open and strutted than in holosteans, which have dense cancellous bones and thick scales, and as a result teleosts can achieve neutral buoyancy with a smaller swimbladder (7% of body volume in freshwater teleosts, compared with over 12% in *Lepisosteus*). Teleost fins are usually very flexible, with fewer rays than holostean fins, and the caudal fin is normally symmetrical. The skull is similar to that of holosteans, but the lower jaw is simplified to consist only of dentary, angular and articular. Internally, teleosts have an elastic bulbus in place of the contractile conus of the holostean heart; gut diverticula are usually more complex; and few teleosts have any kind of spiral valve.

The success of teleosts may be judged by the large number of species, and by the immense numbers of individuals within a species. Teleosts exploit more kinds of living spaces than do any other fishes or aquatic vertebrates. This is partly because they have a wide spectrum of body size, enabling them to fit into many different living spaces. About 10% of teleosts are 10 cm or less in length (Lindsey, 1966), while at least 80% are between 10 cm and 1 m long. The adaptive advantages of these small and medium-sized teleosts are easily seen in an environment such as a coral reef, which provides shelter, feeding and spawning territories suitable for gobioids a few centimetres long on one hand, to lutianids and lethrinids of up to 50 cm on the other. In reef channels, larger predatory carangids prey on the medium-sized teleosts. Another important teleost innovation was the evolution of buoyant eggs, produced by many coastal and deep-sea species, for although mortality is high, wide broadcasting by ocean currents enables marine teleosts to cover and exploit their living spaces.

We have already briefly considered the classification of teleosts (Figure 1.4). Apart from morphological characters, the study of cellular DNA content has more recently illuminated teleost relationships (Hinegardner, 1968; Hinegardner and Rosen, 1972). These studies show that highly specialized genera have less DNA per cell than do more generalized genera in the same group—for example, the paracanthopterygian scopelomorphs average 1.2 picograms DNA per cell, while the more specialized percomorphs average 0.95 pg. The data suggest that fishes with high DNA content are evolutionary conservatives.

4. *Dipnoi*. The six species of living lungfishes (Figure 1.12) were regarded as the most interesting of fishes by Goodrich (1909). The group first appeared in the Devonian, and was widespread in the Palaeozoic. Modern lungfish are in several ways specialized and simplified versions of the earlier, more completely ossified forms. The Australian *Neoceratodus*, however, is similar to the Triassic *Ceratodus*. It is covered with large overlapping scales (thinner, smaller scales with small surface spines are

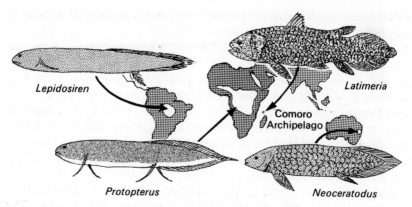

Lepidosiren

Latimeria

Comoro
Archipelago

Protopterus

Neoceratodus

Figure 1.12   Distribution of lungfishes and *Latimeria*. After Greenwood and Norman (1975) and Nelson (1976).

found in the other living genera), the notochord is large and unconstricted and the girdles and fin skeletons are cartilaginous. The paired fins of *Neoceratodus* are lobe-like (sarcopterygian) but in *Lepidosiren* and *Protopterus* they are elongate. Young *Protopterus* 'walk' with their paired fins, very like amphibians. The unpaired fins (found in fossil forms) are absent, but it seems that the symmetrical 'caudal' may have been formed by the union of dorsal and anal fins, since there is no trace in living lungfish development of the heterocercal elasmobranch-like tail seen in the fossil forms.

Lungfishes eat small invertebrates and large amounts of plant material. In modern forms, paired upper and lower toothplates cut and crush the food. The upper jaw is fused to the braincase, and both inhalent and exhalent openings to the nasal chamber lie in the roof of the mouth. Although the exhalent opening was long thought to be equivalent to the internal naris of tetrapods or osteolepid fishes, it is now recognized as simply the homologue of the actinopterygian posterior nasal opening, secondarily displaced into the mouth. The gut is simple and ciliated (there is no stomach or hepatic caecum) but a spiral valve is present. The airbladder is septate and lung-like (see Figure 5.10) paired in *Protopterus* and *Lepidosiren*, but single in *Neoceratodus*, which normally breathes water. A simple valve in the truncus arteriosus allows a partial separation of the pulmonary and systemic circulations, and in this feature, as in the structure of the brain and reproductive system, lungfishes are reminiscent of amphibians. In *Neoceratodus*, the larva is covered with a pattern of ciliated cells which is not seen in other fishes, but is very similar to those in amphibian larvae (Whiting and Bone, 1979).

5. *Crossopterygii*. The living *Latimeria chalumnae* (first caught off Southern Africa in 1938) is known only from a further 75 or so specimens, all from the Comoro archipelago (Figure 1.12). *Latimeria* is the sole survivor of a specialized crossopterygian group which first arose in the mid-Devonian. While it is much larger (80 kg and up to 2.5 m), *Latimeria* is very similar to its forebears in other respects, and evolutionary conservatism has characterized the group. As we would expect by analogy with conservative teleosts, cellular DNA is at a relatively high level (13.2 pg per cell—Thomson *et al.*, 1973).

Latimeria is covered with large overlapping bony spinous scales, and has the typical coelacanth trilobed tail and lobate fins (Figure 1.12). The notochord is very large and unconstricted, and a moveable joint across the middle of the skull allows an increase in gape while feeding. The jaws have numerous small teeth, and there is a large stomach and a spiral valve. The swimbladder is surrounded by fat so that only a narrow lumen remains, connecting with the oesophagus by a ventral opening. The braincase too is lipid-filled and the flesh is oily. These lipid stores reduce *Latimeria*'s density in the absence of swimbladder gas.

Rostral organs in the snout, filled with gelatinous material, may be electroreceptors (Hetherington and Bemis, 1979); in the brain there is the same electroreceptor nucleus that is found in all electroreceptive fishes (Chapter 10). Like elasmobranchiomorphs, *Latimeria* retains urea, which makes the osmolarity of its body fluids close to that of seawater. Despite the lack of obvious intromittent organs in the male, the species seems to be ovoviviparous (see Figure 8.6) with eggs, about the size of tennis balls, like those of deep-sea squaloids.

Although many specimens of *Latimeria* have now been caught, few biochemical or cytological studies have been carried out because the specimens (caught mainly at night on handlines from dugout pirogues) are in poor condition by the time they can be deep-frozen or fixed. A single fish, collected during an expedition in the 1970's, was just alive when it fell into scientists' hands, and is the source of much of our knowledge of *Latimeria* cytology (Locket, 1980).

CHAPTER TWO

LIVING SPACES

2.1 The numbers of fishes and the space available

Living spaces vary enormously in extent. In the south-western deserts of the United States, the isolated pools and hot springs of the Death Valley region hold relict fish populations, including four species of desert pupfishes, *Cyprinodon* spp. One of these, *C. diabolus*, lives in a thermal spring (Devil's Hole) where its single population is not known to exceed 500 (Miller, 1950). At the other extreme, in the largest and most deserted part of the ocean, the bathypelagic zone below 1000 m, the most abundant fishes are minnow-sized black species of the stomiatoid genus *Cyclothone*. The most widely distributed of these, such as *C. microdon* and *C. acclinidens*, range over the three oceans and must be represented by many thousands of millions of fishes.

Estimates based mainly on marine data suggest that the average number of individuals per fish species may be about four thousand million. Horn (1972) suggests that some pelagic species, such as certain anchovies, may attain population levels of about 10^{12} individuals (the figure for rocky shore fishes may be nearer to a million), and that 10^{10} is the average figure for marine fishes. For freshwater fishes he proposes high and low values of 10^{10} and 10^7, thus arriving at values (Table 2.1) for the volume of water (in km^3) available for individual fishes in the sea and fresh waters.

Table 2.1 Volume of water available to individual fishes

Type of species	No. of species	Individuals/species	Volume (km^3) per individual
Marine	11675	10^{10}	1.1×10^{-5}
Freshwater (1)	8275	10^{10}	1.5×10^{-9}
Freshwater (2)	8275	10^7	1.5×10^{-6}

Table 2.2   Productivity and biomass in different ecosystems

Major ecosystem	Net primary productivity (dry wt. in kg per m^2 per year)	Plant biomass (dry wt.: $kg/m^2/yr$)
	(mean value)	(mean value)
Lake and stream	0.5	0.02
Continental shelf	0.35	0.01
Open ocean	0.125	0.003

The degree of isolation and habitat partitioning in fresh waters is obvious from these estimates, as Horn points out, and the total volume available for marine and freshwater species is strikingly different (marine 11 300 km^3; freshwater 15 km^3 if the lower and probably more correct estimate for individual numbers is taken). This disparity may well also be related to the differing levels of productivity and biomass in the two main environments, as seen in Horn's table (Table 2.2) above. As Horn concludes, these values are not in 'great discord' with the estimate that fresh waters hold 10 times as many fishes per unit volume than do marine ones.

2.2   Marine living spaces

2.2.1   The open ocean

In the open ocean beyond the continental shelves there are about 2500 species of fishes, which are subdivided about equally into pelagic and bottom-dwelling species (Cohen, 1970). As the open ocean covers nearly two-thirds of the earth's surface, and has a mean depth of about 4000 metres, it will be plain that a potentially vast volume of water is available per species—about 1 million km^3 in the pelagic zone and half a million km^3 in the combined pelagic and benthic zones, according to Horn (1972).

The pelagic environment is roofed by the euphotic zone, where, during the favourable season in temperate to cold waters, and for almost all the year in subtropical and tropical belts, light is strong enough for photosynthesis by the phytoplankton (Figure 2.1). This zone of primary productivity, which extends down to at least 100 metres in the clearest waters, holds an epipelagic fauna including about 250 species of fishes, most of which are sharks, flying-fishes, scombroids (tunas, billfishes, etc.) and stromateoids (Figure 2.2). Nearly all have their headquarters in warm temperate to tropical regions. Certain species, such as the blue-fin tuna, the blue shark, and isurid sharks, migrate into temperate waters during the

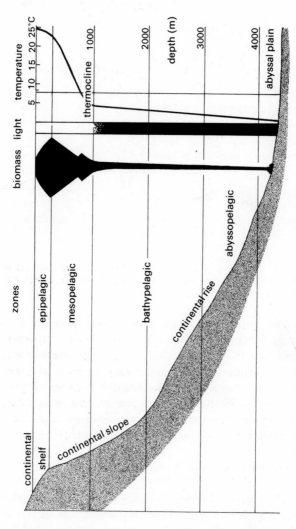

Figure 2.1 Zones of the ocean—note that planktonic biomass increases at the ocean floor. Modified from Marshall (1971).

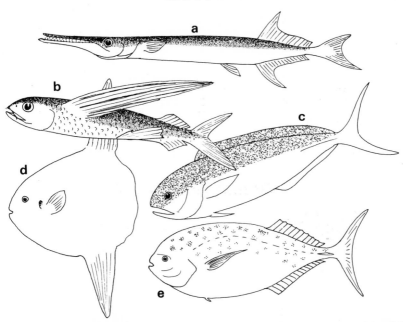

Figure 2.2 Epipelagic fishes (not to same scale). (*a*) Garfish (*Tylosurus*); (*b*) flying-fish (*Exocoetus*); (*c*) dolphin (*Coryphaena*); (*d*) sunfish (*Mola*); (*e*) louvar (*Luvarus*). After Fitch and Lavenburg (1971) and Herre (1928).

productive season. Of these, all except the blue shark (*Prionace*) are warm-blooded (Chapter 3). Indeed, the blue-fin tuna (*Thunnus thynnus*), which migrates from waters off the Bahamas (30°C) to waters of about 6°C in the northern North Atlantic, is able to maintain its body at a relatively constant temperature between these wide ambient limits. Skipjack and yellow-fin tuna, also warm-bodied but whose migrations rarely subject them to water temperatures below 20°C, have no such means of maintaining a constant body temperature (Carey *et al.*, 1971).

2.2.2 Pelagic fishes of the deep sea

Deep-sea pelagic fishes, whether mesopelagic (ca. 200–1000 m) or bathypelagic (below 1000 m) are also predominantly found in (or rather below) the warm temperate to tropical regions of the ocean. There are about 1000 species, most of which (850) are mesopelagic. Two-thirds of the latter are either lanternfishes, Myctophidae (ca. 250 spp.) or stomiatoids. There are also argentinoid fishes (order Salmoniformes), alepisauroids (relatives of lanternfishes), melamphaids (order Beryciformes) and two perciform groups, the Chiasmodontidae (giant swallowers) and trichiuroids (scabbard-fishes, etc.).

2.2.3 Mesopelagic fishes

Though most mesopelagic fishes (Figure 2.3) are small (few that we catch exceed a length of 30 cm), many species, particularly myctophids and various stomiatoids, undertake daily vertical migrations. Those with a gas-filled swimbladder are prominent on the records of echo-sounders; such sound-scattering layers move upward towards sunset and downward before dawn to resume their daytime levels. These migrations take the fish to the more productive surface waters, where analyses of catches made by midwater nets show not only that these are feeding migrations but also that competition is reduced by the existence of vertically segregated feeding levels (Marshall, 1979). Over the tropical and the subtropical belts

Figure 2.3   Mesopelagic fishes (not to same scale). (a) Myctophid (*Electrona*); (b) mycto-phid (*Lampanyctus*); (c) stomiatoid (*Bonapartia*); (d) adult paralepid (*Paralepis*)—these are rarely caught, post-larval and young stages are elongate and thinner; (e) melamphaeid (*Melamphaes*); (f) lancetfish (*Alepisaurus*); (g) alepocephalid (*Xenodermichthys*). After Grey (1964), Rosen (1966), Gibbs and Wilimovsky (1966), and Ebeling and Weed (1973).

of the ocean, a migration from a depth of 400 metres to the surface and back again means that for an hour or so during their ascent and descent fishes experience a rise and fall in temperature of about 10°C. (ca. 10–15°C at 400 m and 20–25°C in the surface waters). Non-migrators, such as the alepisauroids and giant swallowers, have the jaws and capacity (in the form of a distensible gut and body walls) to take large meals, whereas most of the migrators feed on zooplankton (Marshall, 1979).

2.2.4 Bathypelagic fishes

While most of the ca. 150 species of bathypelagic fishes are ceratioid angler-fishes (about 100 spp.), the dominant forms in numbers of individuals are black species of the stomiatoid genus *Cyclothone*. There are also gulper-eels (saccopharyngoids) and whale-fishes (cetomimoids) (Figure 2.4). All live in surroundings that are lit only by fitful sparks of bioluminescence and where temperatures range from about 1–5°C.

Figure 2.4 Bathypelagic fishes (not to same scale). (*a*) Gulper (*Eurypharynx*); (*b*) *Cyclothone braueri*; (*c*) gulper eel (*Saccopharynx*); (*d*) whalefish (*Cetomimus*); (*e*) angler (*Linophryne*) female; (*f*) snipe eel (*Cyema*). After Böhlke (1966), Marshall (1971), and Fitch and Lavenburg (1968).

Although the bathypelagic zone is the most deserted part of the ocean, the fishes are not known to migrate upward in search of food.

Even so, bathypelagic fishes are organized to conform with their food-poor surroundings. Their organ systems, compared to those of their mesopelagic relatives, are much reduced. In particular, the degree of reduction of the skeletal and muscular systems, the heaviest parts of a fish, is such that bathypelagic fishes are near neutral buoyancy even though they are without a gas-filled swimbladder (see Chapter 4). Beside using little energy to maintain their level, they also have the jaws and capacity to acquire potential energy in the form of large opportunist meals. *Cyclothone* spp. exist on a diet ranging from copepods to small fishes, as do female anglerfishes, some of which (e.g. *Melanocetus*) can accommodate fishes two to three times their own length. The entire 'organization' of ceratioid anglerfishes and *Cyclothone* species is also reduced through the development of dwarf males, fused to the female in anglerfishes, thus posing an interesting (and unexamined) immunological problem.

2.2.5 Bottom-dwelling fishes of the deep sea

Most of the bottom-dwelling fishes of the deep ocean, whether by virtue of a gas-filled swimbladder or a large buoyant liver, have become neutrally buoyant, and are thus able to swim easily (and habitually) near the sea floor. Such fishes may be described as benthopelagic. In contrast, the habit of benthic deep-sea fishes, most of which have no special weight reducing system, is to rest on the bottom. As might be expected, bottom-dwellers are more readily observed from deep submersibles or recorded on film than are pelagic species.

The benthopelagic fauna (Figure 2.5) which ranges from upper slope levels (ca. 200 m) to about 8000 m and is cosmopolitan, is dominated in species by rat-tails (Macrouridae) and ophidiid fishes. Both have a gas-filled swimbladder, which in the males of many species is provided with sonic muscles. Deep-sea cods (Moridae) and eels, halosaurs and notacanths also have a swimbladder, but the sharks (squaloids) and chimaeras store oil for static lift (Chapter 4). Most species find food in the benthopelagic zooplankton and among the fauna of benthic invertebrates. In biomass, both these trophic sources decrease exponentially with depth. For instance, at 1000 m the biomass of benthopelagic zooplankton is about 1 % that of the surface water zooplankton, at 5000 m about 0.1 % (Wishner, 1980). Like the deep-sea benthos, the benthopelagic zooplankton is largely dependent on organic matter, ultimately traceable

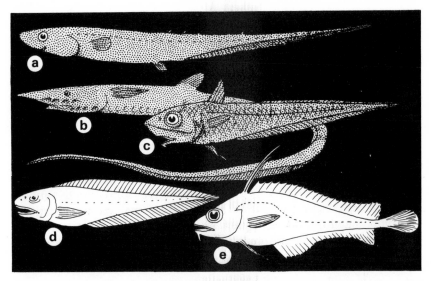

Figure 2.5 Benthopelagic fishes (not to same scale). (*a*) Notacanth (*Polycanthonotus*); (*b*) halosaur (*Aldrovandia*); (*c*) macrourid (*Coelorhynchus*); (*d*) brotulid (*Bassogigas*); (*e*) gadoid (*Lepidion*). After McDowell (1973); Ebeling and Weed (1973).

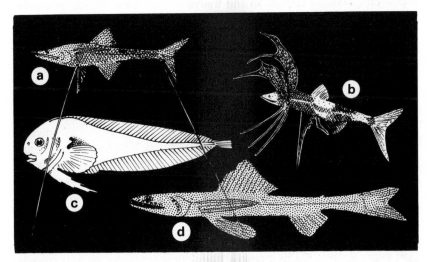

Figure 2.6 Benthic fishes (not to same scale). (*a*) Tripodfish (*Bathypterois grallator*); (*b*) tripodfish (*B. bigelowi*) in fishing attitude facing into current (from photograph by R. Church); (*c*) seasnail (*Careproctus*); (*d*) lizardfish (*Bathysaurus*). After Marshall (1979), Mead (1966), and Marshall (1971).

to the productive surface waters and falling in the water column, particularly in the form of faecal particles and pellets.

The benthic fauna (Figure 2.6) has much the same depth range and overall distribution as the benthopelagic. Round the world, below temperate and tropical seas and from the upper continental slope to abyssal depths (250 to 6000 m), the chlorophthalmids (green-eyes, tripod-fishes, etc.) are dominant forms (Sulak, 1977). Seasnails (Liparidae) and eel-pouts (Zoarcidae) are also important, but both, including many shallow sea representatives, are most diverse from temperate to subpolar regions of the northern hemisphere. All such fishes are negatively buoyant and rest on the deep-sea floor. Tripodfishes (*Bathypterois* spp.) rest on stiff and elongated rays in their pelvic and caudal fins. Raised on this tripod undercarriage, they feed on copepods and other small crustaceans in the benthopelagic zooplankton. Curiously enough, deep-sea skates and rays have large livers, and are probably near neutral buoyancy (Chapter 4), unlike their shallow-water relatives (Bone and Roberts, 1969).

2.2.6 Shallow sea fishes—warm-water species

Of the ten thousand-odd species of shallow-sea fishes, about 8000 live from warm temperate to tropical regions. Most of these warm-water species are associated with coral reefs and atolls, which are spread over an oceanic area of some 68 million square miles. These coral formations, which consist largely of the interlocked and encrusting skeletons of calcareous red algae and reef-building zoantharian corals, flourish best in clear waters where the mean temperatures do not fall below 18°C during the coldest part of the year. In the Indian Ocean and western Pacific Ocean, reefs and atolls are widespread between latitudes 30°N and 30°S. There are also large reefs in the Caribbean and around the West Indies.

Coral fishes are most diverse in the Philippines, New Guinea and Great Barrier reef (Figure 2.7), where respectively 2300, 1950, and 1500 species have been recorded. Only some 500–750 species are known from reefs in Florida, and as Sale (1980) has shown, there is a pronounced cline of diminishing diversity away from the richest central Indo-West Pacific region (Figure 2.8).

Nearly all coral fishes belong to the great superorder Acanthopterygii, and many families of this group, like squirrelfishes (holocentrids), damsel-fishes (pomacentrids), or butterfly- and angelfishes (chaetodontids), are represented wherever corals are found. Coral fishes need to be very manoeuvrable, and often have small mouths adapted for picking their

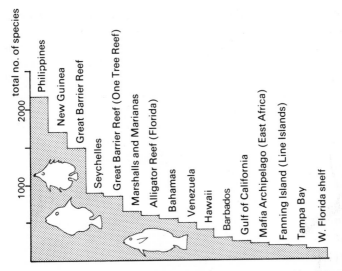

Figure 2.7 Species diversity of coral reef fishes at different reef locations. After Sale (1980).

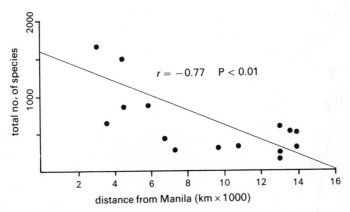

Figure 2.8 Numbers of coral reef species from each of the 16 sites in Figure 2.7, showing geographical change in diversity. After Sale (1980).

food, so it is no accident that many of them have abandoned normal oscillatory swimming, and using their caudal fins as rudders during normal slow swimming, propel themselves along with pectoral fin breast strokes (wrasse, surgeonfishes and parrotfishes); others, like triggerfishes and trumpetfishes, flap dorsal and anal fins (Marshall, 1965). Examples of coral reef fishes are shown in Fig. 2.9.

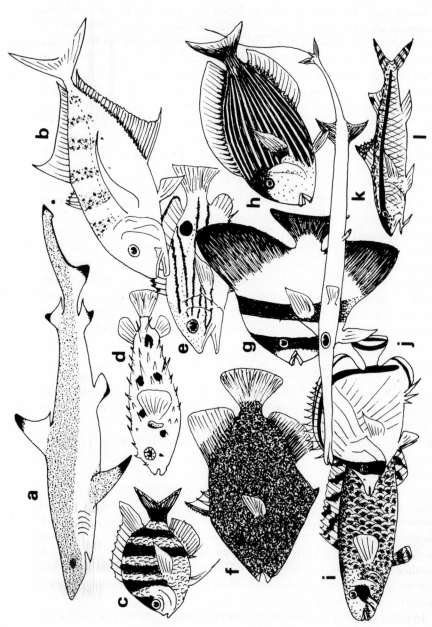

Figure 2.9 Coral reef fishes from the Great Barrier Reef (not to same scale). (*a*) Blacktip shark (*C. melanopterus*); (*b*) jack (*Caranx*); (*c*) damselfish (*Abudefduf*); (*d*) puffer (*Tragulichthys*); (*e*) snapper (*Lutjanus*); (*f*) trigger-fish (*Canthidermis*); (*g*) batfish (*Platax*); (*h*) surgeon-fish (*Ctenochaetus*); (*i*) parrotfish (*Leptoscarus*); (*j*) butterflyfish (*Chaetodon*); (*k*) cornetfish (*Fistularia*); (*l*) goatfish (*Upeneus*). After Marshall (1965).

Many coral fishes crop various kinds of algae, or, like parrotfishes, scrape off living coral tissue containing algae. Herbivorous fishes account for much of the algal biomass produced in tropical reefs. Others, like pufferfishes, boxfishes, gobies and some damselfishes, depend on invertebrate food, whilst some filefishes and butterflyfishes eat coral polyps.

Coral fishes and other warm-water species may be divided into four main faunas: the Indo-West Pacific, the Pacific American (Panamanian), the West Indian and the West African. As we have seen, many spiny-finned teleost families and others are represented in all four faunas. At the species level, such global representation, as we might expect, is much rarer: in each area there are many endemic species. Thus, the Indo-West Pacific fauna, which extends from the Red Sea and the east coast of Africa across the Indian Ocean to the Western Pacific Ocean (to an easterly limit described by a line from the Hawaiian Islands south to the Tuamoto Archipelago) has few species in common with the Pacific American fauna. The Eastern Pacific oceanic barrier between these two faunas, which covers an east-west distance of about 3000 miles, has been crossed by very few species. Those that have crossed include large powerful swimmers such as the tiger-shark, *Galeocerdo cuvieri*, and the spotted eagle-ray, *Aetobatus narinari*, and those with long-lived pelagic larvae like the bone-fish, *Albula*, and six species of moray eels (see also Marshall, 1965, and Briggs, 1974).

2.2.7 Temperate and cold-water fishes

By far the most productive parts of the warm ocean are those that are cooled by upwellings of subsurface water, particularly those associated with eastern boundary currents off California, Peru, and north-western and south-western Africa. Here persistent trade winds blow surface waters away from the coasts, and to replace them, cooler underlying waters rise towards the surface bearing accumulations of nutrient salts. The latter sustain great flowerings of phytoplankton and hence an abundance of small herbivorous forms in the zooplankton. Such mixed plankton supports large schools of clupeid fishes, notably anchovies off Peru and California (since 1952 the California sardine, *Sardinops coeruleus*, has been in decline) and the pilchards *Sardina pilchardus* and *Sardinops sagax* off north-western and south-western Africa respectively. The upwelling regions also support good stocks of demersal fishes, notably different species of hake (*Merluccius*).

In polar and tropical seas there is little change in temperature from one season to the next. Such variation is greatest in so-called temperate waters, which range from cold (sub-polar) to warm temperate in nature. For instance, north of Iceland temperatures may be down to 0°C in February and up to 10°C in August. In the Mediterranean, which except for its subtropical south-easterly parts, may be regarded as a warm temperate sea, the annual range of temperature is from 12°–13°C in February to 20–25°C in August.

The more temperate, subantarctic waters of the Southern Ocean, where temperatures range from about 15° to 5°C, lie between the Antarctic and Subtropical Convergences. The coastal waters, most of which lie over the Chilean and Patagonian shelves, are much less extensive than are the temperate shelf waters in the Northern Hemisphere. The southern fish fauna is also smaller, and over the Patagonian shelf, for instance, most species are members of the Notothenioidi (see Figure 2.10) a suborder of the acanthopterygian order Perciformes. Codfishes (Gadidae), flatfishes (Pleuronectiformes) and herrings (Clupeidae), which are so conspicuous in the fauna of north temperate seas, are poorly represented.

Most of the coastal fishes in Antarctic waters are also notothenioids, but eel-pouts (Zoarcidae) and seasnails (Liparidae) are also found. Virtually all these fishes are quite distinct from their nearest relatives in subantarctic waters. The Antarctic Convergence, which more or less encircles the Southern Ocean, and across which temperatures from north to south drop rapidly by 2° to 3°C, has evidently long been a barrier to the free exchange of coastal fishes in subantarctic and antarctic regions.

For all or most of the year antarctic fishes live at temperatures below 0°C. In the coldest regions around Antarctica temperatures never vary much from −1.9°C. As the body fluids of teleosts freeze at temperatures rather more than 1°C higher than does normal sea water (−1.9°C), the antifreeze adaptions of polar fishes (such as the icefish, Figure 2.10) have attracted attention. Certain arctic and antarctic coastal fishes have supercooled body fluids (like many frost-hardy plants), and they avoid freezing by keeping clear of ice-laden waters. More rigorously adapted antarctic fishes not only contain glycoprotein antifreeze compounds (De Vries, 1971, 1974) but also have aglomerular kidneys (Chapter 6) that preclude the urinary loss of small antifreeze molecules (Dobbs et al., 1974).

The coastal fish fauna of the temperate North Pacific, which consists of about a thousand species, is considerably richer than that of the temperate North Atlantic. In each region scorpion-fishes (Scorpaenidae), sculpins (Cottidae), armed bullheads (Agonidae), seasnails (Liparidae), kelpfishes (Clinidae) eel-blennies (Lumpenidae) and flatfishes are represented by different species, but they are more diverse in the North Pacific, which also contains certain endemic families, such as the surfperches (Embiotocidae) and greenlings (Hexagrammidae). But there are more cod-like fishes (Gadidae) in the North Atlantic, where cod, pollack and haddock are among the species that are common to both the American and European regions.

Lastly, temperate seas, though neither hot nor cold, exhibit, as we saw, a much greater annual range in temperature than do polar and tropical waters. Moreover in tropical and subtropical regions the plankton continues to grow and reproduce throughout the year, whereas in temperate waters there are off-seasons, and temperate fishes, particularly those of the colder regions, thus have much reduced supplies of food during the winter. This drastic change is shown in annual growth rings on the scales, bones and otoliths. The coastal fishes of temperate regions are some of the most adaptable fishes in the sea (Marshall, 1965).

2.3 Estuarine fishes

Sea and land waters overlap and mix in estuarine reaches, which are limited to river mouths in tidal areas. Like other estuarine organisms, the fishes are mainly euryhaline forms that are able to exist in unstable surroundings. Even so, some of the largest estuaries are productive enough to support sizeable fisheries. For instance, along the West African coast the Niger estuary covers about 3400 square miles and consists largely of saline mangrove swamps subject to tidal flooding. At high tide the maximum salinity near the river mouth is about 28%. In the Volta estuary, the salinity varies between mean values of about 15% from August to October and 33% from April to May. The West African shad, *Ethmalosa dorsalis*, flourishes in these surroundings, as do large schools of *Sardinella* in lagoons and river mouths. Though the salinity in the Niger estuary is usually very low, species of marine origin, such as snappers (*Lutianus*), grunts (*Pomadasys*), catfishes (*Tachysurus*), croakers (*Otolithus* and *Sciaena*), threadfins (*Galeoides* and *Pentanemus*), and grey mullet (*Mugil*), predominate in the catches throughout the year (Pillay, 1967).

Beside surviving variable salinities, currents, and food supplies, some estuarine fishes live in muddy surroundings, and certain of their features, especially the prolonged fin rays and diminutive eyes, remind one of deep-sea fishes. As Pillay (1967) points out, the Bombay duck, *Harpadon nehereus* not only looks like a deep-sea fish (Figure 2.10) but is most closely related to a group with deep-sea representatives.

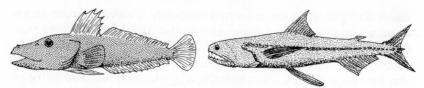

Figure 2.10 Left: icefish (*Chaenocephalus*); right: Bombay duck (*Harpodon*). After Greenwood and Norman (1975), and Jakubowski and Byczkowska-Smyk (1970).

2.4 Freshwater fishes

Most of the eight thousand-odd kinds of freshwater fishes live in land waters that are out of reach of the sea. Even those that tolerate brackish waters must seek fresh waters when ready to spawn. There are a few species, such as carp (*Cyprinus carpio*), roach (*Rutilus rutilus*), and barbel (*Barbus brachycephalus*), that are more adaptable. Populations of these cyprinids in the Caspian and Aral seas produce eggs which will develop at salinities as high as 8 to $10^o/_{oo}$.

Primary freshwater fishes are those that have evolved in fresh waters and have no tolerance for sea water. Cohen (1970) estimates that there are 6650 species, of which about 6200 belong to the carp, characin and catfish order (Ostariophysi). Others include the mormyrids and osteoglossids. Most of the secondary kinds live exclusively in fresh waters, but the main groups (Cichlidae, Cyprinodontidae and Poeciliidae) contain species that occur in fresh or brackish water and have some tolerance for salt water. Diadromous fishes migrate between fresh and salt waters. Of the (*catadromous*) species that leave fresh waters to spawn in the sea, the best known are freshwater eels (*Anguilla*), the genus comprising some 16 species that are widely but patchily distributed in land waters that flow into the three oceans. More species (called *anadromous*) make the reverse migration after periods spent at sea where they feed and grow towards sexual maturity—examples are lampreys, sturgeon, many salmonids, shads (*Alosa, Ilisha* etc.), and alewives (*Pomolobus*). Lastly, certain marine families, for instance the herrings (Clupeidae), drumfishes (Scianenidae) and pufferfishes (Tetraodontidae) contain a number of freshwater representatives.

Freshwater habitats range in level and motion from lotic (washed) kinds in springs, streams and rivers to those in the calmer (lentic) waters of lakes, ponds and swamps. In lotic waters, as may be seen in European river systems, habitats are related to the lie of the land. Starting in the hills, clear and turbulent streams, which even in summer rarely attain temperatures above 10°C, hold brown trout and bullheads (*Cottus*). Further down-slope, where the current is lively and the well-aerated water is clear and cool (summer temperatures rarely exceeding 15°C) is the grayling (or minnow) reach, and salmon, brown trout, dace, chub and stone loach may also have their swim. All but the loach, which hugs the bottom, are active muscular fishes with shapely, fusiform bodies. The slope becomes gentler in the barbel region, where the current is moderate and the waters may be turbid from time to time (summer temperatures are usually above 15°C). Here one may see roach, rudd, bleak and perch, and also dace and chub. Still further downstream, the river widens, runs slowly and is bordered with dense vegetation. The water is often clouded by detritus and is very warm in summer, reaching temperatures of 20°C or more. Near the surface there is adequate dissolved oxygen, but nearer the bottom the supply may be meagre. This is the bream region, where bream, carps, tench, and eels are all able to survive in poorly aerated waters. The first group have deeper bodies and more leisurely ways of moving than the active fishes of the brown trout and grayling reaches.

Ostariophysi. In land waters, as in the ocean, the diversity of fishes is greatest in tropical regions. The freshwater fish faunas of both Africa and South America consist of over 2500 species, and like the faunas of tropical Asia and north temperate lands, are dominated by the Ostariophysi. The characoids, which are the most primitive, live only in South America and Africa and in the former region fill niches that correspond to those exploited by cyprinoids in the other two continents (Figure 2.11). Characoids and catfishes are the dominant fishes in South America. The cyprinoids, which are absent from South America, dominate the fresh waters of tropical Asia as do their relatives in the north temperate zone. Cyprinoids are also prominent in African rivers—most of the communities in the Great Lakes of Africa are dominated by cichlids, which are spiny-finned (percoid) forms. In tropical fresh waters there are about 30 families of catfishes of which 14 are endemic in South America, 8 in Asia and 3 in Africa (Lowe-McConnell, 1975).

Of the three main groups of Ostariophysi, adaptive radiation is most far-reaching in the characoids, which include pelagic herring-like forms, mullet-like forms, gar-like forms, darter or goby-like forms, minnow-like forms and salmonid-like forms. The group also includes characoid flying fishes (Gasteropelecidae) which use their pectoral fins for propulsion while in the air (Weitzmann, 1962).

Compared to a generalized characin with its fixed tooth-bearing jaws,

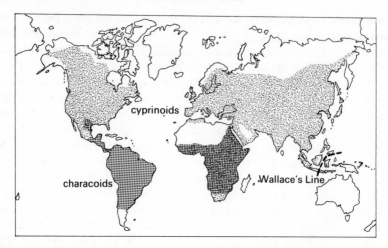

Figure 2.11   The distribution of cyprinids and characoids. After Nelson (1976).

the cyprinids are specialized in their protrusile and edentulous jaws, and are able to inhale small prey, which is passed to the pharyngeal teeth (see Chapter 7). Some species pump in mud, and strain out algae and small animals through the pharyngeal teeth and gill rakers. There are also predatory and herbivorous cyprinids. But as Brittan (1961) contends, the cyprinids are not over-specialized fishes, and this may well have permitted their radiation into all freshwater ecological habitats except mountain torrents and waters very low in oxygen. Catfishes are the ostariophysans best adapted to poorly aerated waters. As Regan (1936) described them, they are carnivores, living on the bottom in stagnant or muddy rivers, with small eyes, and with 6 or 8 elongate barbels used for finding their food (Chapter 9). The olfactory organs are large, and many are found elsewhere on the body apart from on the barbels. Many are nocturnal, fitting into ecological gaps left by cyprinids and characoids, since the latter which have large eyes, are usually diurnal. There are often defensive spines (which can be locked into position) in the pectoral and dorsal fins, and as expected in bottom-dwelling fishes, catfishes are usually flattened and have a reduced swimbladder (Alexander, 1965).

     In contrast, many surface-dwelling fishes (e.g. certain kinds of top-minnows, characids and cyprinids) are small and have an upturned mouth well suited not only to take food caught in the surface film but also to inhale near surface water that in stagnant conditions may be the only source of dissolved oxygen. Mid-stream dwellers are strong swimmers with fusiform bodies. And, as we have seen already, species that live in calm waters among rooted vegetation tend to have deep, laterally compressed bodies.

The overwhelming success of the Ostariophysi in fresh waters (contrasting with the dominance of marine environments by the acanthopterygian teleosts) may well be related not only to their unusually acute hearing (Dijkgraaf, 1960—see Chapter 9), but also to the pheromone signalling system whereby 'fright-substance' is liberated into the water from special epidermal club cells when an ostariophysan is injured. A single cichlid species, *Tilapia macrocephala*, has the same system (Fryer and Iles, 1972), as do also gonorhynchids (Pfeiffer, 1977) which are now grouped with ostariophysans. We do not know why ostariophysans require such good hearing as compared with most other fishes, but as Pfeiffer (1962) suggests, the 'fright-substance' system is particularly suitable for schooling species such as most characins and cyprinids.

2.4.1 Tropical freshwater fishes

The fish communities in tropical fresh waters are reviewed authoritatively by Lowe-McConnell (1975). For instance, in the Zaire river system, which she compares and contrasts with that of the Amazon basin, specific associations of fishes live in the main river, the marginal waters, the inundated forest zone leading into swamps and pools, and in the affluent rivers and streams of various sizes. Immense shoals of plankton-feeding fishes, such as small clupeids (*Microthrissa*) and *Barbus* (Cyprinidae), are found in the main rivers. Some of the mormyrids are also fluviatile, as are the large species of characids and cyprinids. The shallow marginal rivers carry far more fishes than do the open waters. Here there is little current, high temperatures (25°–35°C) and sandy or muddy bottoms may be rich in vegetable detritus. A diversified series of habitats and much fish food attracts many species, which spend all or part of their life in these surroundings. In certain places (e.g. the Yangambi area) there is a basket fishery for many species of mormyrids, characids, catfishes and cichlids (*Tilapia*).

Swamp waters, whether permanent or temporary, are very acid (pH 3.8–5.0), shallow (a few cm to 3 or 4 m in depth) and often overlie a leaf-carpeted bottom of organic mud. The fishes here are specialized for life in deoxygenated waters: they are surface-dwellers, such as top-minnows and those with accessory respiratory organs—lungs in bichirs (*Polypterus*) and lungfishes (*Protopterus*), epibranchial organs in notopterids and

Phractolaemus, arborescent organs in clariid catfishes, and labyrinthine organs in anabantids.

Life in deoxygenated waters has encouraged the evolution of air-breathing organs and jaws properly angled for surface skimming, and has also evoked physiological adaptations in the form of increases in hypoxia tolerance (Hochachka, 1980). Lake Tanganyika is permanently stratified and there is little or no oxygen in the waters below the thermocline, but cichlids are able to live near the bottom (Fryer and Iles, 1972). Quiescent crucian carp and goldfish can also survive in anoxic conditions.

In Asiatic fresh waters there is nothing like the diversity of fishes that live in the Zaire and Amazon basins. Cyprinids are the dominant forms and catfishes are well represented, but there are also numerous species of marine origin. In Thailand, out of some 550 species there are over 200 cyprinids, about 100 catfishes and about 150, including 80 gobioids, of marine provenance. The outstanding feature of the east Asiatic fauna is the great difference in the primary fish faunas on either side of Wallace's Line, which lies down the Makassa Straits between Borneo and Celebes (Figure 2.11). There are over 300 species west of the line in the Bornean fauna, but only two on Celebes to the east, and even these (the labyrinth-fishes, *Anabas testudineus* and *Channa striatus*) have probably been introduced.

In Australia the only true freshwater fishes may be the lungfish, *Neoceratodus forsteri*, and the osteoglossid, *Scleropages leichhardti* (found also in Southern New Guinea). The others, such as silversides (Atherinidae), gobies, galaxiid 'trout' and 'smelt' (retropinnids), may have come from the sea.

CHAPTER THREE

SWIMMING

3.1 The problem of analysis

Fishes swim with such apparent ease that it is hard for us to realize (until we dive ourselves) just how difficult it is to move rapidly through water, and how well adapted fishes are for locomotion in a dense medium. It is not surprising that our own underwater designs are often very similar in shape to fishes, for the density and viscosity of the medium dictate the same streamlined form: the solid of least resistance, which the great pioneer aerodynamicist Cayley took from the body of a trout (Figure 3.1), looks just like a modern submarine hull. Yet there is an essential difference between the fish design and our own which makes the analysis of fish swimming a difficult and intriguing problem.

Submarines are rigid vehicles which use rotating propellers to provide the thrust required to overcome the drag on their hulls. It is possible, therefore, to make rigid models and to test them in water tunnels to collect data which can be used to estimate the drag they incur and the power

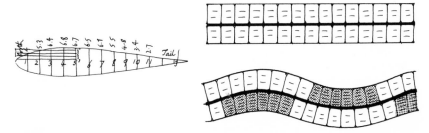

Figure 3.1   Left: Cayley's drawing of a trout; right: schematic diagram illustrating bending of a body with flexible backbone (thick black line) by segmented muscle blocks (contracted, stippled). After Gibbs-Smith (1962).

39

needed to drive them along. Data from such tests and from experience with actual submarines and aircraft form the basis for theoretical analyses of the hydrodynamics of rigid bodies. Unfortunately for biologists, fishes are usually far from rigid, and they generate thrust not with propellers, but in quite a different way, usually involving change in shape of the body as they pass through the water.

Various fishes have adopted special methods of locomotion (not all for use in water) in which they use paired and unpaired fins to row, flap, or undulate themselves along, but the great majority of fishes propel themselves by a combination of two processes—the backward passage of transverse waves along the body, and lateral movements of the caudal fin. In such fish as scombroids, thrust is generated only by the caudal fin, and the body flexes rather little, whereas in eels, or in dogfish, there are large-amplitude body waves passing towards the tail. Most fishes fall somewhere between these two extremes, sending smaller amplitude waves down the body which terminates in a substantial caudal fin, as in gadoids or salmonids. In general, the fastest swimming fishes rely on the oscillation of an high aspect ratio caudal fin, attached to the body by a narrow caudal peduncle; slowly-swimming fish undulate through the water like eels, and as we shall see, this is a less efficient process.

This flexible method of thrust generation makes it impossible to apply directly hydrodynamic data from the performance of rigid bodies to the analysis of fish swimming. We could, for example, make a model of an eel, or stiffen an actual eel in some way, and measure its drag in a water tunnel, but we should not expect that the drag we measure would be the same as that which must be overcome by the swimming fish, nor that the surface flow patterns around the model would be even remotely like those around the living fish. Although the hydrodynamics of fish swimming are thus more complex, and less well understood than those of rigid bodies, good progress has been made in the past decade, both in obtaining kinematic and respirometric data from swimming fishes set up in water tunnels, and in working out suitable mathematical models of fish swimming. We can now make reasonable estimates (at least for small fishes) of the power needed for swimming, and of the efficiency of the process.

In this chapter we shall look first at the machinery which fishes use to move their tails and bodies; next see how these movements generate propulsive thrust; and finally, look into the efficiency of swimming, and at the adaptations which some fishes have devised to reduce the cost of swimming—some of these remarkably resemble devices found on modern aircraft.

3.2 The muscular machinery

3.2.1 Myotomal structure

The backward passage of body waves and the lateral oscillation of the tail both result from the operation of the segmented axial musculature, divided up into myotomes by the myoseptal connective-tissue partitions on to which the muscle fibres insert. The myotomal muscle fibres run roughly parallel to the long axis of the fish, and because the notochord or backbone is incompressible, contraction of some or all of the myotomes on one side of the body will result in lateral movements (Figure 3.1). Stated simply like this, it seems easy to understand how the myotomal apparatus works, and the main problem would seem to be the organization of the control system in the spinal cord ensuring that the appropriate link in the chain of successive myotomes along the body contracted at the correct point in the swimming cycle.

Certainly, this is an interesting problem (considered in Chapter 9) but a closer look at the myotomes of any real fish raises other questions; there are mechanical subtleties in the myotomal layout which are not yet fully understood.

Firstly, no fish myotomes are simple rectangular blocks like those of Figure 3.1. Not only do the myocommata slope obliquely forwards and backwards from the myotomal surface just under the skin, to their insertion on to the backbone (Figure 3.2), so that transverse sections always cut across several myotomes, but the outer borders of the myotomes are folded. In amphioxus (Figure 3.2) they are simply folded once, into a series of V-shapes, but in lampreys and higher fishes (Figure 3.2) the folding is rather more complex, and gives a shallow W-shape. The phylogenetic increase in complexity from amphioxus to teleosts seen in Figure 3.2 is paralleled in teleost ontogeny, where the myotomes form first as V-shaped blocks before folding further to give the adult W-shape. It is not easy to make 2-dimensional diagrams of what are after all, rather complicated 3-dimensional structures, and the attempt of Figure 3.2 is no substitute for looking at the myotomes themselves. Probably the best approach is to lightly boil a fair-sized fish so that the myotomes can be separated, and then to try and model myotomes for oneself out of plasticine. Within these complicated myotomes, only the most superficial muscle fibres are actually arranged parallel to the long axis of the fish. Those deeper in the myotomes are arranged in a kind of spiral pattern so that each lies at a different angle to the long axis, even up to 30°. Figure 3.3 shows the arrangement diagrammatically. At first sight this arrangement

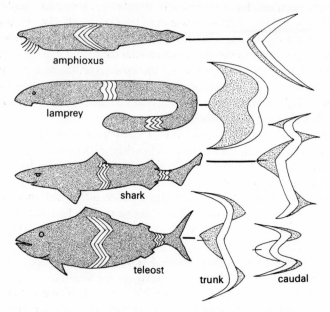

Figure 3.2 Myotomal form in different fishes. Single myotomes are seen at right, their outer surface unstippled. After Nursall (1956).

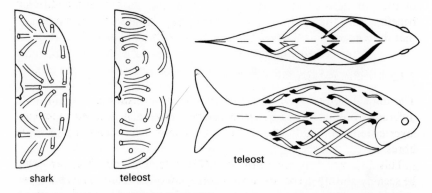

Figure 3.3 Left: thick transverse slices across a shark and a teleost showing orientation of myotomal muscle fibres. Note tendons in shark (horizontal lines). Right: dorsal and lateral views of typical teleost showing course of muscle fibres in successive myotomes along the body. The helices shown were obtained by taking the origin of one muscle fibre from the point at which the muscle fibre in the myotome next anterior inserts on to the common myoseptum, and so on along the fish. After Alexander (1969).

seems very peculiar, for a completely parallel arrangement might seem more what is wanted, but as Alexander (1969) showed, the spiral design allows each muscle fibre within the myotome to contract at the same speed, in this way operating most economically. Alexander and Kashin's analyses were notable, for they offered the first functional explanation for a puzzling feature of myotomal design, and what is more, hint at the reason for the complex folding of the myotomes, probably arranged to allow optimum muscle fibre packing in a wide myotome.

The myosepta are composed of a meshwork of inextensible collagen fibres, which can be deformed, but not stretched (like a sheet of paper), and they are in teleosts often stiffened to limit their deformation by vertebral ribs and by intermuscular bones. It is not yet clear exactly how these stiffening elements work (they are absent from all sharks) but it seems probable that they allow only lateral movements when the myotomal musculature contracts, and serve to reduce the general flexibility of the body. In oceanic scombroids, and other high-speed teleosts, the posterior myotomes are elongate in their lateral folding, thus giving rise to tendons which operate the caudal peduncle.

The advantage of this arrangement is clear: it is the same as that which determines that the leg muscles of fast-running terrestrial animals lie within the body, rather than across the joints in the limbs. The mass and inertia of the caudal peduncle is kept as low as possible by the development of these tendons so that the rate of tail beat may be increased and the width of the peduncle be reduced for hydrodynamic reasons. An additional benefit is that the inertia of the anterior part of the body is increased, so that the tendency for the tail to oscillate the body is reduced.

3.2.2 Myotomal muscle fibres

The muscle fibres of the myotomes in almost all fishes are of two quite different kinds, easily distinguished by their colour. A thin red superficial layer covers the major part of the myotome which is made up of white fibres.

This division of the myotomes into two sorts of muscle fibres can easily be seen by cutting across the post-anal region of the body. The lateral red strip of muscle is darker than is the remainder of the myotomal muscle. A closer examination of the red and white parts of the myotome shows that the muscle fibres in each are quite different from the other (Table 3.1, Figure 3.4).

Table 3.1 shows that the red muscle strip is designed for aerobic

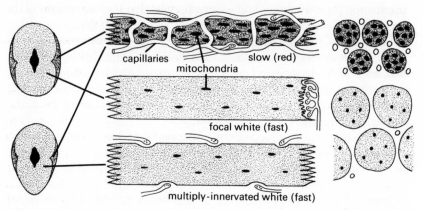

Figure 3.4 Summary diagram showing structure and innervation pattern of red and white myotomal muscle fibres in fishes. Note that multiply-innervated white fibres are found only in higher teleosts (see text).

operation, whilst the poorly vascularized, large white fibres, poor in mitochondria, seem plainly designed for anaerobic operation. It seems as though the fish has two very different motor systems in the myotomes, operating on different fuels (or at least utilizing different routes for ATP

Table 3.1   A comparison between the fast and slow muscle fibres in fish myotomes (Bone, 1978).

Slow	Fast
Smaller diameter (20–50% of fast fibres)	Large diameter (may be more than 300 μm)
Well vascularized	Poorly vascularized
Usually abundant myoglobin, red	No myoglobin, usually white
Abundant large mitochondria	Few smaller mitochondria
Oxidative enzyme systems	Enzymes of anaerobic glycolysis
Lower activity of Ca^{2+}-activated myosin ATPase	High activity of enzyme
Little low molecular wt. Ca^{2+}-binding protein	Rich in low molecular wt. Ca^{2+}-binding protein
Lipid and glycogen stores	Glycogen store, usually little lipid
Sarcotubular system lower volume than in fast fibres	Relatively larger sarcotubular system
Distributed cholinergic innervation	Focal or distributed cholinergic innervation
No propagated muscle action potentials	Propagated action potentials; may not always occur in multiply-innervated fibres
Long-lasting contractions evoked by depolarizing agents	Brief contractions evoked by depolarizing agents

production from the fuels), and we might guess that the two regions of the myotome are used by the fish for different kinds of swimming.

Several lines of approach support this guess, but the most direct is simply to insert wire or needle electrodes into the muscles of swimming fish, and record electrical activity under different conditions of swimming.

Electromyograms (EMG's) recorded in this way from herring swimming in a tunnel respirometer, or from 'spinal' dogfish (which are particularly convenient since they will continue to swim although the brain has been removed) show that the red portion of the myotome is active when the fish swims slowly, and that under these cruising conditions, no activity is found in the white portion of the myotome. Only when the fish is swimming in rapid bursts are muscle action potentials visible from the white muscle (Figure 3.5). Although the fish can continue to use the red part of the myotome for long periods and some fishes (such as mackerel or pelagic sharks) cruise continually the whole of their lives using red muscle fibres, the operation of the white part of the myotome is severely limited in dogfish and herring, being exhausted after 1–2 min of continuous operation. This is not such a drawback as might appear, for it is very rare for any fish to swim rapidly for more than a few strokes of the tail—it then glides into slower swimming. In fact, 2 min continuous operation in dogfish would allow the fish to travel about 600 m.

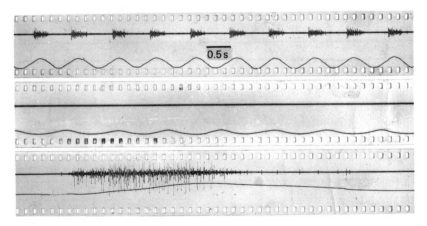

Figure 3.5 Electromyographic records from swimming spinal dogfish. Upper line in each case represents electrical activity from muscle fibres, lower line the swimming movements of the fish. Top: electrode tips in red muscle; middle: electrode tips in white muscle; bottom, same as middle, but white muscle activated by pinching tail of fish. Bottom trace at higher recording speed than the others, to show separate muscle action potentials from white fibres.

3.2.3 The cost of speed

Most of the drag which the fish must overcome as it swims along results from the viscosity of the water (water particles tend to become attached to the surface of the fish, forming a boundary layer), and is called skin friction drag (D_{sf}). We shall look at the different sorts of drag in the next section, but for the moment, all we need to know is that skin friction drag is proportional to the *square* of the speed at which the fish is swimming. At constant speed, drag must be equal to the thrust the fish produces (or it would slow down or speed up), and so the power required from the muscles to generate this thrust will be proportional to the *cube* of its swimming speed.

This means that speed in the water is extremely costly in terms of the power needed from the muscles. Doubling the speed from, say, the $25\,\mathrm{cm\,s^{-1}}$ cruising speed of a dogfish, to $50\,\mathrm{cm\,s^{-1}}$ requires not twice, but eight times the cruising power (other things being equal). It is small wonder that in most fish the white part of the myotome is much bigger than the red portion, and that the white muscle fibres are highly specialized for maximum power production. Ultimately, the power produced by a given amount of muscle depends on the actin-myosin interaction, so that any modifications which will allow more myofilaments in a given volume will increase the power output. Thus the white muscle fibres are large (up to $300\,\mu m$ diameter), have very few mitochondria interrupting the myofilament array, and have few muscle capillaries occupying space which could be filled by muscle fibres. These specializations for maximum power output naturally imply that the white fibres operate anaerobically, and as we find from their enzyme profiles and from studies on metabolite depletion, they operate by anaerobic glycolysis. This is a relatively inefficient way of gaining ATP for driving the actin-myosin interaction, and the ion pumps of the fibres, since this route provides only 3 mols of ATP per glycosan unit. Fast swimming is indeed an expensive process. But all designs are a compromise between conflicting requirements, and the fish accepts a relatively low chemical efficiency for its high-speed swimming machinery in return for a highly-efficient system that is used for continuous operation.

One difficulty arising from the use of anaerobic glycolysis for short bursts of speed is that large amounts of lactate are produced. Castellani and Somero (1981) have shown that a good correlation exists (as it should) between muscle buffering capacity and activity in different fish.

The red muscle fibres make up in most fish only 5–15% of the total myotome, and operate by aerobic glycolysis (yielding 38 mols of ATP per glycosan unit) or by aerobic lipolysis (yielding over 100 mols of ATP per mol); plainly a much more efficient use of the fuel. These processes require oxygen and so red fibres have a rich capillary bed, contain myoglobin for internal oxygen transport, and contain vast numbers of large mitochondria—in some scombroids, half the volume of the red fibres consists of mitochondria. Of course, this means that the power produced per unit of muscle volume will be much less than that for the white fibres, but at cruising speeds what is needed is efficiency and economy of operation, not maximum power output.

3.2.4 The two motor systems and terrestrial animals

The manifold differences between the red and white fibres, summarized in Figure 3.4 and in Table 3.1, emphasize the difference in design of the two for their two different roles, and reinforce the idea that fishes have two separate, independent motor systems in their myotomes, used under different swimming conditions. The arrangement is analogous to that of modern military jet aircraft, where economy in the cruise condition contrasts with the vastly increased power obtained by the occasional emergency use of reheat; extravagant in fuel, but acceptable for short bursts.

The dual motor system of the fish myotome is very different to the arrangement of muscle fibres in higher terrestrial animals. In water there is little penalty for carrying around a large mass of white muscle that is only of occasional use, but on land the animal is not buoyed up, and such a system would incur a severe weight penalty. So our muscles, and those of other terrestrial animals, are arranged in a different way, and our muscle fibres are all more similar to each other than are those of the red and white portions of the fish myotome.

3.2.5 Cruising speed and red muscle

The proportion of red fibres in the myotomes gives a good estimate of the importance of sustained cruising in the life-style of the fish, so that, for example, salmon and spur dogs (*Squalus*) have a relatively larger amount of red muscle than do dogfish or similar sluggish fishes. But it is not possible simply to increase the mass of red muscle (to increase cruising speed) beyond a certain limit, for as we have seen, it requires abundant oxygen to operate, and this has to be acquired by the gills. Probably 15 % of the total

muscle is the greatest mass of red muscle that can be adequately supplied
by the gills, unless (as in scombroids) there are special adaptations to
increase oxygen acquisition and transport (Chapter 5).

3.2.6 Warm red muscle

In some fishes, such as the more advanced scombroids and the big isurid
sharks, body temperature is above ambient (Carey *et al.*, 1971), and the red
muscles are run at relatively high temperatures. There is no problem in
generating heat (it is an inevitable byproduct of muscular contraction,
which is why we shiver when we are chilled), but at the gills of fish the
blood is so intimately in contact with the surrounding water that it must be
at ambient temperature (the gill is like a radiator); special arrangements
have to be devised to maintain high muscle temperatures. Warm fishes use
countercurrent heat exchangers, by making the warm blood leaving the
red muscles pass through a capillary rete mirabile where it runs close to the

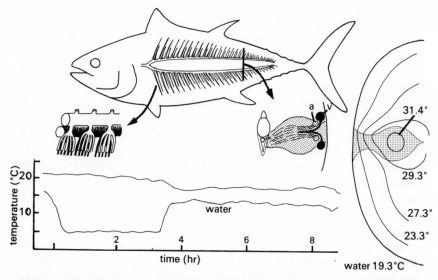

Figure 3.6 Organization and function of the retial thermoregulation system in the bluefin
tuna (*Thunnus thynnus*). Upper: bluefin showing lateral vessels (details of branches to left,
position of lateral vessels and retial system in red muscle to right. a, artery; v, vein). Right:
thermal profile obtained with thermistor probes, red muscle stippled. Bottom: records of
water temperature (lower line) and stomach temperature (upper) obtained by telemetry from
free-swimming bluefin, showing independence of body temperature from changes in water
temperature. After Carey *et al.* (1971), Gibbs and Colette (1967).

cool oxygenated blood, and so heat is exchanged to the incoming blood and the muscle is able to maintain its elevated temperature (Carey and Teal, 1966). The anatomy of this ingenious system is seen in Figure 3.6.

Similar countercurrent rete mirabile are found in the flippers of cetaceans, and in the swimbladder gas generating system (Chapter 4), both places where the added capillary resistance is outweighed by the advantages of heat and gas retention. Although warmblooded fishes have long been known, it is still not clear *why* it should be advantageous to be warmblooded, (see Neill *et al.*, 1976). One possibility is that more power could be extracted from a given volume of muscle if it is 'run' at a higher temperature, but the increase in white muscle volume that this would allow (remembering that the red muscle volume is probably limited by its oxygen requirement) seems but a small gain for the complexity of the retial system. Perhaps more important is the possibility that the fish is able to operate its cruising musculature at a more or less constant temperature, in waters of different ambient temperature. That is, like mammals, the advanced warmblooded fish can maintain a more or less constant 'milieu intérieure', and can adapt its enzyme systems to operate at a single most efficient temperature.

3.2.7 *Functional overlap between the two motor systems*

This brief resumé of the dual motor system in the fish myotome has suggested that the two systems are indeed quite separate and employed during different kinds of swimming. No biologist will be surprised to learn that although such a generalization is broadly true, there are some exceptions. All the available experimental evidence suggests that the two motor systems are separate in sharks, lungfish, lampreys, amphioxus, *Amia*, and in some primitive teleosts like clupeids or eels. But in the more advanced teleosts (that is in most fish) the fast motor system is probably used not only during bursts of fast swimming, but also during slow sustained cruising.

In carp, for example, if we record from their red and white muscles whilst they are swimming at different speeds in a tunnel respirometer (Figure 3.7) we find low-level electrical activity in the white muscle even at the lowest speeds at which we can persuade the fish to swim. At higher speeds, bursts of action potentials of the usual kind are seen from the white muscles just as in dogfish or trout. The reason for this difference is not yet fully understood (Bone *et al.*, 1978; Bone, 1978), but seems to be related to the unusual (distributed) innervation pattern of the fast fibres in these teleosts

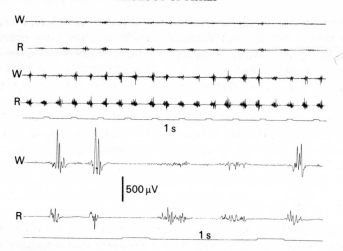

Figure 3.7 Electromyogram from red (R) and white (W) regions of myotome in carp swum at different speeds in a tunnel respirator respirometer. Note that even at the slowest speed at which the fish could be persuaded to swim (0.75 body lengths s^{-1}, top) low-level electrical activity is seen in the white zone, increasing as swimming speed increases to 1.76 b.l. s^{-1}. Bottom record shows spike-like potentials recorded from white zone at 2.0 b.l. s^{-1}. Time marks seconds. After Bone *et al.* (1978).

(Figure 3.4). It does mean that some use can be made of the otherwise inactive white muscle during sustained cruising. Just how white fibres whose whole design seems to be organized for anaerobic operation can be used during sustained swimming remains a mystery, for an oxygen debt is not accumulated by these fish during long periods of sustained cruising.

3.3 Generation of thrust

3.3.1 Caudal fin oscillations

The earliest method of generating thrust was certainly that still used by lampreys and eels, that is, the propagation of waves back down the body, but it is simplest to begin with the much more advanced caudal fin oscillations used by tuna and other fast fishes, which have virtually separated the propulsive machinery from the thrust-generating apparatus: they are much closer to our own rigid machines. Tuna are essentially well-streamlined masses of myotomal muscle, attached to a rigid high aspect ratio lunate foil by a narrow neck (see Figure 3.6). The myotomal muscles are attached to tendons, which pass across a flexible joint in this narrow caudal peduncle, so that as they contract the foil is oscillated from side to side across this joint, acting much like a propeller (though an oscillating rather than a rotating one). Because it oscillates rather than rotates, the caudal foil operates alternatively with the same surface forming the upper and lower surface of the foil, and since it comes to a halt, as it were, at the end of each stroke, the thrust produced is oscillatory, though smoothed by operation at high frequency, up to the remarkable frequency of 10 tail beats per s. How does this oscillatory foil generate thrust, and how is it specially adapted to do so? Here we are on firm ground, for the process is the same as that by which aeroplane wings generate lift, or propellers thrust.

3.3.2 Circulation, lift and thrust

The forces which act on a foil as it passes through a fluid arise from the displacement of fluid by the foil; they are of two kinds. Frictional forces (resulting from viscous effects) can for the moment be neglected, since they are only indirectly concerned. Lift or thrust forces can only result from a net difference in the pressure of the fluid on either side of the foil. As an aeroplane wing passes through the air, it is both sucked upwards from above, and pushed up from below, for there are higher pressures below the wing and lower pressures above (see Figure 3.12).

How do these pressure differences arise? The relationship between pressure and velocity in the flow of an incompressible fluid is given by

$$P + \tfrac{1}{2}\rho V^2 = \text{constant}$$

where P = static pressure; ρ = density, and V = velocity. The quantity $\tfrac{1}{2}\rho V^2$ (often abbreviated to q) is the *dynamic pressure*, the force experienced if one puts one's hand out of the window of a moving car, and it is this which is the fundamental source of aerodynamic or hydrodynamic forces. As we shall see, the dynamic pressure appears in formulae for the calculation of thrust, lift and drag.

Since the density of the fluid does not change at the speeds at which fish swim, clearly the pressure distribution which one can measure over a foil in a fluid (Figure 3.8) must result from differences in the velocity of flow over the two surfaces, and the velocity is higher above a lifting foil than below it. If small particles like polystyrene beads are distributed in water, and a foil is moved through the water, the displacement of the water caused by the passage of the foil can be observed by taking high-speed cine-films. Let us follow the displacement of two particles A and B from their initial positions at rest before the foil arrives, until their final positions after it has

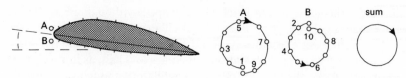

Figure 3.8 Circulation and lift. Two particles A and B are caused to change position as the foil passes through the fluid. Right: the successive positions of the two particles (numbering as on left). The sum of their movements is a net clockwise circulation producing lift.

passed. We find that they follow curved paths of opposite sense (Figure 3.8), that is, they are forced to *circulate* by the passage of the foil. Since the displacement of particle A above the foil is greater in the same time than is that of B below it, the algebraic sum of the two results in a net clockwise rotation of *circulation*, and because it is in this sense, the relative velocity of fluid above the foil will be greater than that below it, and lift will be generated. We can see that the stronger the circulation the greater the lift.

3.3.3 The reduction of drag associated with lift and thrust

The circulation over the main part of the foil continues downstream in a series of tip vortices at the tips of the foil as it moves through the fluid, giving rise to an upwash just behind the back of the tip of the foil, and a downwash across the span, decreasing to the root of the foil. This process requires energy, and the energy that has to be expended to maintain the tip vortices is a component of the drag the lifting foil incurs as it passes along, called vortex drag (D_v).

It is not hard to guess that as the tip vortices depend on the circulation, the greater the circulation, the greater the amount of vortex drag which is incurred. Like aeroplanes, fish take advantage of two possibilities to reduce this drag. First, since a given lift or thrust can be produced by a weak circulation over a long span, or by a strong circulation over a short span, doubling the span can halve the vortex drag incurred, and for this reason, the caudal fins of fast-swimming fishes are long and thin, as are the lifting pectoral fins of fast fishes like tuna, which are denser than water (Figure 3.9). Their aspect ratio, AR ($\text{span}^2/\text{fin area}$) is as high as 6; as high as can be arranged without sacrificing structural strength. Secondly, it can be shown that for a given aspect ratio, vortex drag is least if the spanwise lift distribution is elliptical; this can be conveniently achieved by making the plan form of the foil elliptical, and this is the shape both of the caudal foils of scombroids, and of the lifting pectoral fins of fast sharks and many large fishes like marlin or swordfish (Figure 3.10). High AR tail fins of this

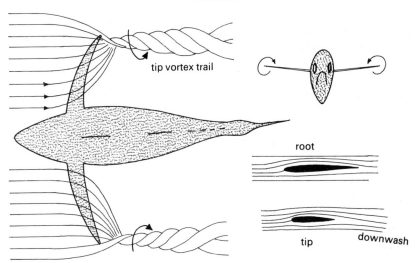

Figure 3.9 Flow patterns around a gliding skipjack (*Katsuwonus*). Tip vortices cause vortex of lift-associated drag. Lower right: flow over root and near tip of pectoral fins showing downwash near tips.

kind, and the similar lifting pectoral fins, are designed to minimize vortex drag, and are suitable for high-speed cruising. But they are not very suitable for rapid manoeuvring, or for rapid accelerations, and this is why fishes which need these capabilities have caudal fins that are much broader and of lower aspect ratio. To provide thrust or (in the case of an aeroplane or dense fish) lift, the foil has to be operated at an angle of attack to the

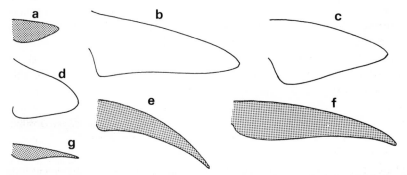

Figure 3.10 The lifting pectoral foils of different fishes (teleosts stippled). (*a*) Mackerel (*Scomber*); (*b*) *Carcharhinus longimanus*; (*c*) basking shark (*Cetorhinus*); (*d*) dogfish (*Scyliorhinus*); (*e*) carangid (*Trachurus*); (*f*) longfin tuna (*Thunnus alalunga*); (*g*) swordfish (*Xiphias*).

fluid, or the circulation of fluid above and below the foil will be equal, and no lift or thrust will result. At small angles of attack, fluid will flow around the foil in an ordered way, but if the angle of attack is increased above a certain limit, the flow around the upper part of the foil will separate from it, leading to loss of lift and greatly increased drag: the foil will stall. A long thin foil of high AR is much more prone to stall as the angle of attack is increased than is a short wide foil of low AR, and it is when accelerating or manoeuvring that the angle of attack of the caudal fin is greatest. Thus the scombroid or albatross elongate foil is not suitable for salmonids or for pheasants, which require rapid acceleration from rest.

In many respects, then, the thrust-generating foils of tuna show similar design features to aeroplane or bird wings, or propellers. Most fishes are less highly modified for rapid swimming than are tuna, and gain part or all of the thrust they need by a rather different process.

3.3.4 Body waves

If we look at successive frames of an eel swimming forwards (Figure 3.11) we see that it passes transverse waves of increasing amplitude down its body, and these pass backwards at a higher speed than the fish swims forwards.

Eels are very flexible, and there is just about a complete wavelength of this locomotor wave along the body, but in less flexible fish, such as trout, only about half a wavelength is seen along the body. Although we could consider this process in a similar way to the operation of the tuna tail, that is, in terms of circulation theory, it is more appropriate, as Lighthill (1971) showed, to think of it in terms of the forces 'felt' by the water masses next to

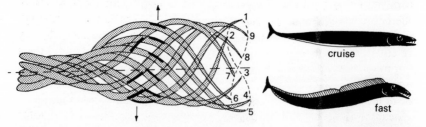

Figure 3.11 Left: drawings from frames (1–9) of film of young *Anguilla*, superimposed to show short sections of the body (thick lines) as planes inclined to axis of progression when the eel moves forwards. The oblique dashed line is the axis of progression, the tail tip describes a figure-eight pattern. After Gray (1933). Right: the two swimming patterns of the trichiurid *Aphanopus*.

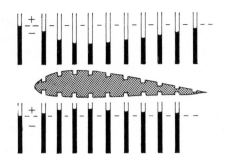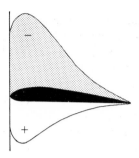

Figure 3.12 Pressure distribution around a foil at a positive angle of attack in a wind or water tunnel. Left: manometers connected to openings in upper and lower surfaces show pressures above and below ambient (dotted line). Right: typical pressure distribution around such a lifting foil.

the body surface. These are the forces which result from the inertia of the water next to the fish, and are proportional to the rate of change of relative velocity of the surface of the animal, with respect to the water next to it. The result of the passage of the locomotor wave backwards along the fish is to increase the momentum of water passing backwards (as it is pushed by the inclined 'planes' seen in Figure 3.11), and it is the rate of shedding of this momentum into the wake which is proportional to the thrust.

The eel is a relatively inefficient swimmer (although it travels enormously long distances to spawn) because the body is more or less round in section, and so the water affected by its passage through it is of relatively low mass. By deepening the body to give a flat cross-section like many carangids, or the leptocephalus larva (see Figure 8.7) the virtual mass of water which is given momentum by the swimming movements is much increased, and greater thrust can be produced. Many fish which swim in this way deepen the section of the body with median fins of different kinds (retractable in teleosts, but not in sharks) and although these median fins play an important role in stabilizing the fish in roll, they are equally important in thrust generation. It is not necessary to have a continuous fin. An interrupted series of fins (like those of many gadoids) works equally well, provided their spacing is appropriate, and this gives rise to less drag.

Steady swimming with a caudal foil, or by the passage of locomotor waves along the body, or by a combination of the two, has been most studied, and reasonable estimates of the efficiency of the process have been obtained. But less is known about acceleration, and as Webb (1977) has shown, the optimum body shapes for steady performance are different from those required for fast-start performance. Some fishes actually adopt different shapes for the two requirements. The deep-water scabbard fish, for example, has an elongate body terminating in a small caudal fin of fairly high aspect ratio. It swims slowly by keeping the body rigid and oscillating this foil. To accelerate rapidly, it unfurls its long median dorsal fin, and passes locomotor waves along the body (Figure 3.11). Amputation experiments on salmonids have shown that for steady swimming, caudal fin removal has little effect on swimming speed. Most of the thrust generated must be provided by the locomotor waves passing along the body. But removal of fins, particularly the caudal fin, reduces acceleration performance, and the conclusion is that the wide, deep caudal fins of many fish are adaptations to rapid acceleration rather than for steady swimming as are the deep bodies of many fish of this kind (such as carp). For rapid swimming, increase in body depth increases the drag, and specialists in rapid swimming, like flying fish or garfish, have thin elongate bodies.

3.4 Drag

3.4.1 Boundary layers and the different drag components

The thrust that the fish generates is at constant speed, equal to the drag forces resisting forward motion: how do these arise, and how can they be minimized by fish? Obviously, drag-reduction is greatly advantageous for fish, since it will mean that for a given power output, they can travel faster, or can use less power at a given speed. Since water is viscous, it tends to stick to the surface of the fish, and a thin boundary layer forms, in which there is a steep velocity gradient from the still water layer next to the fish, carried along with it, to the water outside the boundary layer. This velocity gradient means that there is a large shear stress which results in drag.

The properties of the boundary layer depend upon the ratio between viscous and inertial forces acting on the fish, given by the dimensionless Reynolds number (Re). This is defined as

$$\frac{\text{length} \times \text{velocity} \times \text{density of medium}}{\text{viscosity of medium}}.$$

or LV/v, where v is the kinematic viscosity of the medium, viscosity$/\rho$, approximately 0.01 in water.

The Reynolds number gives us an immediate idea of the relative importance of viscous and inertial forces, and in practice, at Reynolds numbers below 10^5 viscous forces predominate and the boundary layer on a flat plate will be laminar. At higher Reynolds numbers than this, say over

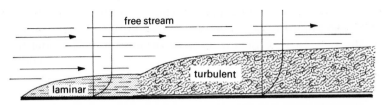

Figure 3.13 Development of boundary layer on a flat plate in a fluid flowing from left to right. The anterior thinner laminar layer has a steeper velocity gradient than the posterior turbulent layer.

10^6, inertial forces are more important, and flow within the boundary layer will normally be turbulent. The two kinds of flow within the boundary layer are seen in Figure 3.13.

In a flat plate of sufficient length, we shall find that near its front upstream edge, the boundary layer will be laminar, but as we pass downstream it becomes thicker, and as L (and hence Re) increase, it changes into a turbulent boundary layer. The point of transition from laminar to turbulent depends much on surface irregularities, for since energy to maintain the laminar layer can only enter by diffusion across the layers within it, it is rather sensitive to any adverse pressure gradients which may be produced by surface irregularities.

The skin friction drag on our flat plate is given by $D_{sf} = \frac{1}{2}\rho V^2 A Cf$, where ρ = density; V = velocity of flow; A = wetted area; and Cf is a drag coefficient.

The value for Cf is much lower for laminar than for turbulent boundary layers. But because laminar boundary layers are relatively unstable, as we have seen, they tend to separate from the surface, and give rise to a turbulent wake behind the object in the flow. A large wake means a large pressure difference between nose and tail, and thus a large pressure drag, so that separation is to be avoided as far as possible. Abrupt changes of contour and awkward excrescences will cause the flow to separate; fishes, however, are smoothly shaped and well streamlined—many have elaborate transparent eye fairings, for example, or fold their fins into slots when they are not required. Fish can very greatly reduce the form drag component of pressure drag. Indeed, a mackerel, say, tested in a water tunnel as a dead rigid object is so well streamlined that its drag is very similar to that of a flat plate of the same wetted area: there is almost no pressure drag. In the swimming fish, however, some part of the vortices trailing behind in the wake result from the generation of thrust (and of lift in dense fishes), so that although fishes minimize vortex drag as far as possible, this second component of pressure drag is an inevitable consequence of thrust and lift generation. The total drag that a fish or an aeroplane has to overcome as it goes along thus has two different components:

total drag: skin friction drag + pressure drag (form drag + vortex drag).

Streamlining and attention to increasing the aspect ratio of the lifting and thrust-generating foils will minimize pressure drag, which is usually of much less consequence than skin friction drag. What can be done to minimize skin friction drag?

3.4.2 Drag-reducing mechanisms

It is important to realize first, as we saw at the outset, that skin friction drag on most fishes is not likely to be the same as that for a flat plate of equivalent wetted area. There are good theoretical and experimental grounds for supposing that it will be considerably greater, by a factor of 2–5 times, for those fishes which make substantial lateral body movements (Lighthill, 1971). For example, the power output estimated from oxygen consumption measurements on goldfish swimming at constant speed was found to be 3.6 times that needed to overcome the drag on a rigid model of the same wetted area (Smit et al., 1971). It is obviously important for fast swimming fishes to reduce the lateral movements of the body, and hence this augmented skin friction drag, and this is just what they do; only the slower swimmers have large amplitude body movements. The great majority of fishes are sufficiently small that even at burst speeds, they do not exceed a Reynolds number of 10^6, so that they do not need to have special devices for reducing skin friction drag, since boundary layer flow will be laminar. But larger and faster fishes swim at Reynolds numbers above this, where inertial forces are more important, and the boundary layer will be turbulent, except in the anterior region. Since the drag coefficient for a turbulent boundary layer is much higher than that for a laminar layer, anything the fish can do to maintain a laminar layer, or to delay transition to a turbulent layer, will reduce its drag. Furthermore, separation of the boundary layer from the surface (with the penalty of pressure drag added to skin friction drag) is more likely at higher Reynolds numbers when the boundary layer is thicker.

Trachypterid and stromateid fishes have a remarkable 'porous' integument which has a canal system filled with sea water just under the smooth outer surface; and it seems likely that this curious arrangement is designed to maintain a laminar boundary layer, by a process of distributed dynamic damping. Walters (1963), who first examined trachypterid fishes such as *Desmoderma*, suggested that small adverse pressures in the boundary layer which might otherwise lead to transition or to separation, would be damped by fluid sinking into the pores and being transmitted to regions of lower pressure. This mechanism has not been tested experimentally, and unfortunately, is not used on our own designs, so that we do not know whether these very specialized integuments could work in this way.

Other fishes have a different kind of skin, which seems to be specialized for maintaining a turbulent boundary layer, and preventing separation. In this case, the system is better understood. The castor oil fish, *Ruvettus*, has a regular pattern of very sharp pointed scales all over its body, which project about 1.0 mm above the body surface (Bone, 1972). These will project through the boundary layer, and entrain vortices from the free stream in the way shown in Figure 3.14. This input into the boundary layer will help to stabilize it and prevent separation. Similar devices are found on

Figure 3.14 The specialized integument of the castor oil fish *Ruvettus*. Upper left: ctenoid scales (white dots) form a regular pattern over the surface. Upper right: transverse section of body showing sub-dermal spaces (black). Middle: stereogram of integument showing ctenoid scales (CT) and sub-dermal spaces (S). Anterior to left. Lower left: assumed operation of integument injecting momentum into boundary layer (see text). Lower right: vortex generators on aircraft wing. After Bone (1972).

some aeroplane wings. One problem is that they will *increase* drag at low speeds, and so they are' only found on fish which can accept reduced efficiency at cruising speeds in order to obtain a high burst speed. Perhaps the ctenoid scales of most advanced teleosts work as vortex generators, but experimental data are still needed here.

As well as vortex generators, *Ruvettus* also has another system for drag reduction by injecting fluid into the boundary layer. The bases of the ctenoid scales form pillars supporting the outer skin away from the connective tissue layer over the muscles, and so the fish has a system of spaces under its skin (Figure 3.14). These are full of sea water, and communicate with the boundary layer by openings facing obliquely backwards. As the body oscillates, water will alternately be sucked into, and ejected from, these sub-cutaneous spaces, and so *Ruvettus* presumably once again stabilizes its boundary layer. *Ruvettus* certainly has the most complex integument of any fish yet examined, although other fishes, like the salp-eating stromateoid *Tetragonurus*, have analogous injection systems. Similar fluid injection into the boundary layer is common in aircraft practice, as for instance in 'blown' flaps and control surfaces, fed from the jet engines.

Quite a different approach to the problem of decreasing drag and thus increasing range (or saving fuel) arises from the increased drag incurred by the lateral body movements over that incurred by the same shape when it is relatively rigid. Any fish which is denser than water could alternate periods of swimming obliquely upwards with periods of gliding down-wards, during which its body would be more rigid. Weihs (1974) has shown that if a tuna or mackerel adopted this swimming pattern, it would achieve a significant advantage (up to about 20%) over steady continuous swimming in one horizontal plane although we do not know whether migrating tuna do swim in such a glide-power cycle. Fishes which have larger lateral body movements than scombroids could save over 50% of the energy required for continuous swimming. The same arguments apply to birds, which are easier to observe than fish, and it is not surprising that many small birds like woodpeckers or finches oscillate up and down as they fly along.

3.5 Efficiency

Efficiency is essentially measured by the ratio of input to output, and there are various measures of efficiency depending on what input/output properties we want to consider. We could compare the swimming fish to our own machines, such as motor cars, which gain power (as does the swimming musculature) by converting chemical to mechanical energy, and then have to transmit this power to the road to drive themselves along. Engineers define two principal kinds of efficiency for such machines.

1. Mechanical efficiency (η_m): $\dfrac{\text{brake horsepower}}{\text{indicated horsepower}} \times 100$.

This is the ratio of the power developed at the pistons vs. the power available at the output shaft after frictional losses in bearings, gear trains, etc. In practice, the mechanical efficiency of a motor car is usually around 80–85%.

2. Thermal efficiency (η_t): $\dfrac{\text{work done}}{\text{heat supplied}} \times 100$.

This is the ratio of one kind of energy, heat energy, supplied during the combustion of fuel in the cylinders, to the useful energy obtained in the form of work done by the engine driving the wheels. Diesel engines have thermal efficiencies around 40%, petrol engines rather less.

Overall efficiency, H, will be the product of these two, so that the overall efficiency of a well-designed diesel lorry will be something like 30%. How do fish compare with this? In the case of the fish, N may be considered as the product of the efficiency of the caudal propeller (usually called η_p), and that of the conversion of chemical to mechanical energy in the muscles (η_m). Values for H during sustained aerobic swimming by salmonids work out between 5% at low speeds to around 20–22% at the maximum speeds sustainable. Calculations for a different biological machine, man, during sustained activity, give similar values.

Obviously, if we can measure H from oxygen consumption during swimming, and if we knew either η_m or η_p, we could obtain both. But unfortunately η_m has not been measured for any fish owing to difficulties with the way in which myotomal muscles insert into compliant myocommata. Still, fish muscles are fairly similar to those of other animals in which the measurements enabling η_m to be determined have been made, and by substituting these values, Webb (1971) calculated that propeller efficiency in trout rose to a maximum around 75% at a swimming speed of 2 body lengths per s. It seems very probable that scombroids and other high-speed fish, swimming by oscillating lunate tail foils, will have a higher η_p than this, and thus that overall efficiency H will be higher than for trout. So fish compare quite favourably with our transport designs, and indeed, with ourselves!

3.6 Speed

Fish vary widely in overall size (from larvae only a few millimetres long to the 12 m whale-shark), and the speed of only a very few within this range has been studied, but it seems generally true that fish smaller than half a metre long are capable of maximum speeds of about 10 times their body length per second. It does not seem likely that larger fish are capable of such a performance (which would give the whale-shark a maximum speed of 422 kph!), although they can travel at relatively high speeds. Accurate measurements of the speed of large fish are hard to come by, but measurements have been made up to 42 kph of line run-out on hooked tuna about 1 m long. A novel approach to the estimate of maximum swimming speeds has recently been made by Wardle (1975) who reasoned that the maximum contraction velocity of the white muscle could be related to swimming speed, since swimming speed is closely related to tail beat frequency. His observations have only so far been made on the muscles of fish under a metre long, but they give maximum swimming speeds in good agreement with those observed. In the case of tuna (Wardle and Videler, 1979) where maximum contraction speed of the white muscle is similar to that of other fish, but whose maximum swimming speeds seem well established as much higher, it is possible that this is achieved by reducing body flexure and maintaining a constant angle of attack of the caudal tail as it sweeps across. Wardle's concept of stride length is an interesting one, but the case of tuna requires further investigation.

CHAPTER FOUR

BUOYANCY

4.1 Dynamic lift

We saw in the previous chapter that some fishes are denser than the water in which they swim, and have to generate dynamic lift mainly by using their outspread pectorals as lifting foils. This process inevitably generates drag, which makes a significant contribution to the total drag and energy requirement of such fishes. For example, a mackerel weighing around 25 g in sea water has to generate some 1.2×10^4 N of dynamic lift during level swimming—mackerel effectively are climbing a 1 in 15 hill all their lives! But increased energy expenditure is not the only drawback to keeping aloft in the water by generating dynamic lift. Dense fishes must maintain a minimum forward cruising speed to generate enough lift to prevent themselves sinking, so they cannot hover or swim backwards. The swimming rhythms of spinal sharks (35–40 tail beats $\min^{-1}$ in spinal *Scyliorhinus*) are apparently related to this minimum swimming speed. For benthic fishes that rest on the bottom and only swim occasionally, dynamic lift generation is appropriate, but for other fishes, a better solution would seem to be to store light materials to provide *static* lift (as do airships and submarines) and thus avoid the drawbacks of generating dynamic lift.

4.2 Static lift

Most of the materials making up a fish are denser than water. For example, much of the body consists of locomotor muscle (density 1050–1060 kg m^{-3} in common marine fishes). Skeletal tissues loaded with heavy mineral salts are correspondingly denser (2040 kg m^{-3} for typical teleost bones). Nevertheless, water is so much denser than air that even without special

stores of low density materials, fishes are not so much denser than the
water in which they swim. They have the option (denied to birds) of storing
sufficient low-density material that they can make themselves the same
density as the water; they can achieve neutral buoyancy and need expend
no muscular energy to keep station in the water.

Fishes use two quite different materials to provide static lift. Gas is
efficient in giving lift, since its density is very low, and most teleosts possess
gas-filled swimbladders. Usually the swimbladder is around 5% of the
volume of marine fishes, and around 7% in freshwater fishes, providing
enough lift for neutral buoyancy. Fish swimbladders cannot significantly
resist changing in volume as the fish swims up and down in the water and
the ambient pressure changes, indeed, the swimbladders of almost all
teleosts obey Boyle's law perfectly (Figure 4.1). So if a fish with a gas-filled
swimbladder is to remain neutrally buoyant at different depths, it must
secrete or absorb gas to keep the swimbladder at constant volume as the
ambient pressure changes. To regulate the *mass* of gas within the
swimbladder in this way requires complex mechanisms of great physiological interest.

Fats and oils are also used by fishes as sources of static lift. These are
much less efficient and much bulkier for a given amount of lift. However,
they have the great advantage that the lift provided varies little with depth,
because changes in ambient pressure have relatively little effect on the

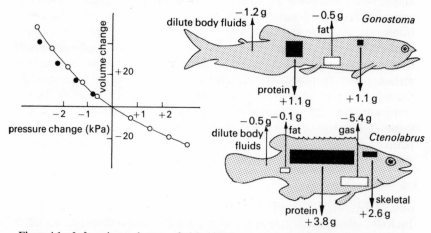

Figure 4.1   Left: volume changes of airbubble (open circles) and fish swimbladder (filled
circles) as ambient pressure changes. Right: buoyancy balance sheets for a mesopelagic fish
(*Gonostoma*) and a shallow-water acanthopterygian (*Ctenolabrus*). After Alexander (1966)
and Denton and Marshall (1960).

volume of the fat or oil: if a fish using lipid as a source of static lift is neutrally buoyant at the surface of the sea, it will be close to neutral buoyancy at the sea bed, even at considerable depths.

In the short term then, lipid provides fewer problems of buoyancy regulation than does gas, but in the longer term, difficulties arise, because the lipids stored to provide lift may have other functions as well. For example, sharks which store oil in the liver and muscles may have to draw on this store as a fuel for continuous swimming, or as a food reserve for developing embryos or for the adult. Where lipid is the sole source of static lift, there will certainly be great complexities in the regulation of lipid metabolism, and the fish will find it difficult to adjust its density rapidly to cope with the short term density changes resulting from feeding and parturition.

Fish using gas as a source of static lift can afford to support the heavy components of their bodies without difficulty, but where lipids are used, the safety margin is not so great, and a number of fish using lipid for static lift can only achieve neutral buoyancy by reducing the dense components of their bodies and so reducing the lift they require. Their skeletons are reduced and poorly calcified, and they even reduce the amount of protein in the muscles, which are weak and watery. Figure 4.1 shows the buoyancy 'balance sheets' (Denton and Marshall, 1958) of a typical shallow-water teleost using gas as a source of static lift, and of a deep-sea form using lipid, and in addition, reducing its dense components. The extreme cases of this 'evasive' solution of the buoyancy problem are the deep-sea predatory fishes which live in midwater, such as *Saccopharynx* and the deep-sea angler fishes. As Denton and Marshall (1958) pointed out, they are little more than floating traps, with weak skeletons and watery muscles; they attract their prey with luminous lures and need not pursue it actively like the fish of the upper ocean. To some extent reduction in density of body components has taken place during teleost evolution, from the thick-scaled 'ganoid' fishes to the thin-scaled clupeids of today. Many fishes also have planktonic larvae whose dense components are reduced, and which have spaces for low density dilute body fluids (Figure 8.8).

4.3 Lipid as a source of static lift

In four unrelated fish groups (and perhaps in others) enough lift is provided by stored lipids to achieve neutral buoyancy. Interestingly enough, the lipids stored are not biochemically similar, but they are all of particularly low density and so most efficient in providing lift. Fish lipids vary in density from around $930\,\mathrm{Kg\,m^{-3}}$ (cod liver oil, and the oils of dogfish livers) to the wax esters of myctophids, gempylids, and *Latimeria* (densities around $860\,\mathrm{Kg\,m^{-3}}$), and the hydrocarbons of some sharks ($860\,\mathrm{Kg\,m^{-3}}$). The difference in specific gravity between cod liver oil, and the wax esters or shark hydrocarbons may not seem very striking, but 1 g of the less dense oil will provide 0.1675 g of lift in sea water whose density is $1027.5\,\mathrm{Kg\,m^{-3}}$, whereas a gram of the denser oil will only provide about half as much lift (0.0975 g). It is thus well worthwhile for the fish to store these lighter lipids rather than the more common triglyceride metabolic reserves, as all the fish that achieve neutral buoyancy using lipids actually do.

4.3.1 Squalene

The first fish found to use lipid to achieve neutral buoyancy were the deep-sea squaloid sharks studied by Corner *et al.* (1969) (Figure 4.2). Unlike the spur dog of the same family, these fish live near the bottom in deep water, and the habitat is one where the family has successfully diversified, for there are many genera known. All share two striking characteristics: very large livers (making them grossly corpulent) and very small pectoral fins. In most animals, including ourselves, the liver is around 4–6 % of the total weight, but in these fish, it may be more than $\frac{1}{4}$ of the total weight, because it contains an enormous amount of pale yellow oil. On this oil the fish literally float—when their livers are removed they sink. In all species examined, the liver oil is of low density (870–$880\,\mathrm{Kg\,m^{-3}}$) because it is mainly composed of the hydrocarbon squalene. Squalene, which was first isolated from such sharks, is formed by the condensation of isoprene units on the pathway leading to cholesterol. It has the low density of $860\,\mathrm{Kg\,m^{-3}}$, and so is admirably suited to provide static lift.

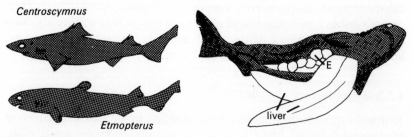

Centroscymnus

Etmopterus

Figure 4.2   Neutrally buoyant deep-sea squaloid sharks. Right: female *Etmopterus* opened to show huge liver and large eggs (E). From sharks caught on a deep sea long line by G. R. Forster from RV *Sarsia*.

We do not yet know how sharks regulate their liver lipids so as to balance their weight in water, but it appears from experiments upon *Squalus* that the fine adjustments required for neutral buoyancy may depend not on changing amounts of squalene, but rather on varying the other less abundant lipid constituents of the liver oil. By attaching weights to *Squalus*, Malins and Barone (1970) found that the fish responded by increasing the amount of low-density alkoxydiglycerides in the liver oil, compared with the control fishes which had larger amounts of the denser triglycerides.

Deep-sea squaloids bear live young and have very large eggs (about the size of a billiard ball in a shark a metre long); ingeniously enough the eggs contain squalene and are neutrally buoyant themselves, so that pregnancy does not increase the density of the mother.

It is obvious that these fish need only have relatively small pectoral fins; they are used only during manoeuvring. Although the tails in all genera are markedly heterocercal, they evidently do not generate lift (see Chapter 3). Why do deep-sea squaloids need to be neutrally buoyant if they live on the sea bed? The answer is that they hover just off the bottom (as we know from cine films taken by deep cameras), unlike the dense bottom-dwelling elasmobranchs of shallow water, which rest *on* the bottom and are invariably very dense, like the dogfish.

Deepwater Holocephali evidently live in a similar way to the deep-sea squaloids, and like them, are close to neutral buoyancy, although they only manage this by virtue of reduction of dense components, and have poorly calcified skeletons. Their liver oil consists largely of squalene and this is also the main source of static lift in some teleosts. Eulachon (*Thaleichthys*), for instance, contain 20% of lipid by weight. Eulachon do not have a gas-filled swimbladder, and during their spawning migration, squalene forms a higher proportion of the total lipid than at other times; it seems that the fish metabolizes reserve triglycerides during these migrations and becomes denser, so that in this case, lipid metabolism is not sufficiently well-regulated to cope with buoyancy and metabolic demands upon the total lipid pool, and at the same time, maintain neutral buoyancy.

4.3.2 Wax esters

A little squalene is also found in the living coelacanth *Latimeria*, but most of the massive amounts of lipid stored are wax esters (Nevenzel *et al.*, 1969). These make up 30% of the wet weight of the ventral musculature and over 60% of the wet weight of the swimbladder (which contains only lipid, no gas), and in the pericardial and pericranial tissues, the percentage is even higher. We know very little about the habits of *Latimeria*, but its fins certainly seem unsuited to provide dynamic lift, and the only specimen examined alive floated at the surface, so that it seems safe to assume that it is neutrally buoyant in life.

Wax esters are probably stored to provide lift in many families of mesopelagic teleosts, but the only ones examined so far have been the gempylids and a few of the numerous kinds of myctophids. The gempylid *Ruvettus* (which has the remarkable integument described in Chapter 3) is loaded with low-density oil (density $870\,\mathrm{Kg\,m^{-3}}$) which has purgative properties, hence the common name of castor oil fish. The cranial bones have been modified as oil tanks, and are the least dense tissues of the body. This large (1 m) predatory fish ranges from depths of 15 m to over 500 m, and is very close to neutral buoyancy, feeding on smaller fishes which undertake diurnal vertical migrations.

The most remarkable of these are the myctophids, some undertaking daily a double journey of 500 m up and down in the water column.

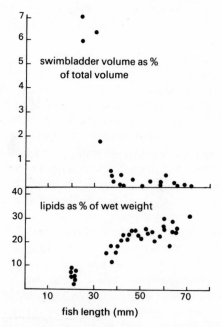

Figure 4.3 Changes in swimbladder gas content and lipid investment during growth of the myctophid *Diaphus theta*. After Butler and Pearcy (1972).

Although all myctophids have gas-filled swimbladders as larvae, in many species the swimbladder shrinks and becomes invested with more and more lipid as the fish gets older, until no gas remains. As in *Latimeria*, lift from gas is replaced by lift from lipid, which may eventually make up 15% of the wet weight of the fish. As we should expect, the species with a high lipid content are neutrally buoyant, and store low-density wax esters. Why should many myctophids have abandoned gas as a source of static lift? Although the evidence is only circumstantial, it seems a reasonable guess that it is because there are difficulties in regulating buoyancy over a wide depth range when gas is the source of lift. Those species which as adults have much low density lipid and are neutrally buoyant, have a greater depth range than the juveniles, which still have gas in the swimbladder (Figure 4.3), and similarly, species which have gas-filled swimbladders as adults undergo less extensive vertical migrations than those which rely on lipid only.

On the whole, the evidence is that lipid storage for static lift is a secondary phenomenon in myctophids, and that the most 'advanced' species in the family are those which have abandoned gas altogether as adults; it is interesting that amongst the family there are fishes showing all stages in this changeover adapted to different life-styles (Figure 4.4).

4.3.3 Insufficient static lift for neutral buoyancy

So far, we have considered fishes of different groups which store low-density lipids to attain neutral buoyancy. What of fish which use lipid for static lift, but are not neutrally buoyant? Many teleosts are certainly in this category. Mackerel vary in density at different times of year because they have more or less lipid stored in the muscles; such a system is evidently a primitive one, for lipid is not set aside, as it were, to maintain a constant density. Rather, the lipid store is at the mercy of metabolic demand, and reduction in density, valuable as it must be, is simply a side effect of the storage of lipid for metabolic purposes.

We know more about elasmobranchs—Baldridge (1969) and Bone and Roberts (1969) showed that a wide range of elasmobranchs, including large pelagic sharks, are each of a characteristic density, some being fairly close to neutral buoyancy, others being very dense. Liver lipid is an important factor in determining the density of some species, but in most, it is the density of other tissues (amount of fat in the white muscle, mineralization of the skeleton) which determines the density of the species. Not only do different elasmobranchs store different amounts of lipid in

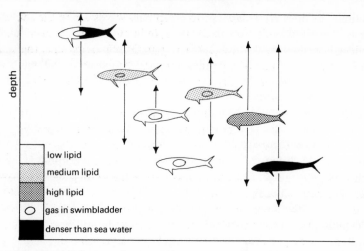

Figure 4.4   Functional types of myctophids. Vertical arrows indicate known or assumed extent of vertical migration. After Bone (1973).

their livers, but the stored oil is least dense in the least dense species. It seems that elasmobranchs have managed to set apart the lipid used for density regulation, whether in the muscles or liver, from metabolic stores, so that the fish can regulate its density to a characteristic figure, whatever metabolic demands are made. How this is done remains to be discovered.

At the beginning of this section, the advantages of neutral buoyancy were extolled, and it seems odd to find fishes using static lift to maintain a fixed density below that of neutral buoyancy. The regulatory mechanisms are there, and at first sight, it seems a harder problem to maintain a constant density (say around 1.03) than to regulate to neutral buoyancy, and what is more, dynamic lift must still be generated, with the penalties that this incurs. However, it is not hard to see why many pelagic sharks use lipid to reduce their density, but not so far as to become neutrally buoyant.

Shark fins vary a good deal in flexibility, and in their stiffening by skeletal elements, but the basic design, unlike that of teleosts, makes them impossible to retract: they are not of varying geometry. If we bear this in mind, we can see how considerations of dynamic lift generation lead to different densities in different species, depending on their mode of life. Reduction in density by lipid storage means that less lift need be generated in level swimming, so that the shark can either reduce the size of its pectoral fins, or it can cruise more slowly without stalling. Since sharks use their pectoral fins for manoeuvring (to turn, change in pitch, and even sometimes, as does *Heterodontus*, to creep backwards along the sea bed), there is a limit to reduction in fin size. Reduction in fin area is wholly beneficial (since it reduces drag), but to remain manoeuvrable, the shark must have pectoral fins of a certain size, and below this, no benefit will be gained by reducing density, apart from a slight reduction in vortex drag.

We saw earlier that vortex drag is proportional to $1/V^2$, so that this will be most important at low swimming speeds, and unimportant at high speeds. We should expect from these considerations that sharks which swim fast would be of reduced density, but not neutrally buoyant; those which swim very slowly (for example as part of the feeding pattern) would be close to neutral buoyancy; and those living on the bottom, would be very dense. Looking into the matter, this is just what we find. The huge whale-shark and basking shark which sieve plankton while swimming very slowly, are close to neutral buoyancy; the fast pelagic tiger and blue sharks are of reduced density, but are not so close to neutral buoyancy; and the bottom-living dogfish and rays of shallow water have no special arrangements for static lift.

Since teleosts have differently designed fins, which *can* vary their geometry, the arguments above do not apply, and we should not expect to find any teleosts using lipid to reduce their density to a constant figure not close to neutral buoyancy; so far, none have been found.

4.4 Gas as a source of static lift

Because fish swimbladders obey Boyle's law (Figure 4.1), to obtain constant lift as they swim up and down in the water, fish are obliged not only to store gas, but also to secrete and absorb it. In the sea, the ambient pressure increases by 1 atmosphere (101.3 kPa) for every 10 m increase in depth, so that a fish living at the surface (with the swimbladder gas at atmospheric pressure) will have the volume (and lift) of the swimbladder halved if it descends to 10 m, and vice versa. It is hardly surprising that the regulatory mechanisms cannot cope effectively with such large changes, and that many surface-dwelling pelagic fish have either lost the swim-bladder altogether (tunas and *Scomber*) or have a reduced swimbladder (*Scomberomorus*). Gas is not a suitable source of static lift for this mode of life, unless the fish remains strictly at the surface, as do hemirhamphids and flying fishes.

At greater depths, however, changes in ambient pressure have relatively less severe effects; a fish changing its depth by 10 m, at say, 400 m, will only

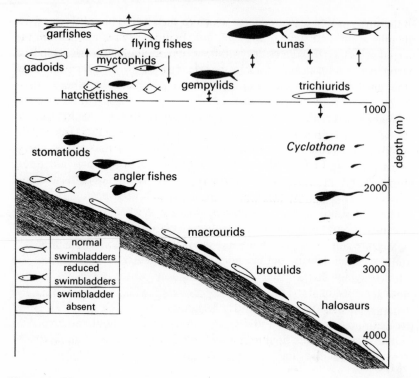

Figure 4.5 Distribution of fishes with and without swimbladders in the oceans. After Marshall (1960).

alter the volume of its swimbladder by 1/40th, and the same applies with greater force, the greater the depth at which the fish lives.

The distribution of fish with gas-filled swimbladders in the sea has been examined by Marshall (1960) who has found that they occur down to the greatest depths (Figure 4.5), but are absent from the upper layers unless they live just at the surface. The deeper the fish lives, the less significant are changes in volume with depth, but the more formidable are the problems of gas secretion and retention. Yet fish with gas-filled swimbladders have been caught at great depths. The macrourid *Nematonurus armatus* has several times been caught below 4000 m, on one occasion at 4500 m. At this depth the ambient pressure is 450 atmospheres (45.6×10^3 kPa).

The swimbladder gas in deep-sea fish is mainly oxygen (for reasons discussed later) so that the partial pressure of oxygen (pO_2) in the swimbladder of such a fish will be over 400 atmospheres (40.5×10^3 kPa).

In the surrounding seawater, from which oxygen has to be collected across the gills and then transferred to the swimbladder via the circulation, the partial pressure of oxygen is several orders of magnitude less than this. The partial pressure of a gas in a gas mixture is the pressure it would exert if the other gases were removed. Thus pO_2 in a 50:50 oxygen–nitrogen mixture at 2.0×10^3 kPa will be 1×10^3 kPa. The partial pressure of a gas throughout a liquid is the same as that of the gas in the gas mixture at the gas–liquid interface. In the sea, therefore, the gases dissolved are in equilibrium with those of the atmosphere at the surface, and at any depth, the partial pressure of oxygen will be around 20 kPa, unless it is lowered by utilization by organisms. At 4500 m then, the partial pressure gradient across the swimbladder wall will be at least 2000:1.

The secretion of gas against this gradient, and its retention in the swimbladder, may justly be regarded as one of the greatest feats of biological engineering; for obvious reasons these processes have not been studied directly in deep-sea fish, but there is no reason to suppose that the solutions to the formidable problems of the deep sea are different in *kind* to those accessible for examination in shallow-water fishes. For one thing, the structural peculiarities of the swimbladder are essentially similar at whatever depth the fish studied normally lives.

4.4.1 Swimbladder structure

Swimbladders vary in structure, partly because they perform other functions than hydrostatic ones (Chapters 5 and 9); in this section we are concerned only with the structures that have evolved to enable the fish to cope with the problems of using gas as a source of static lift. In ontogeny, the swimbladder arises as a pouch from the roof of the foregut, and in many of the more primitive fish, such as elopomorphs or eels, the connection with the gut is retained as an open tube in the adult. These *physostomatous* fish are able to take in air at the surface, and to void it as they ascend: whether the swimbladder originally evolved as an hydrostatic organ or (more probably) as a respiratory organ, there is hardly any doubt that this is the primitive condition, as shown by its taxonomic distribution amongst teleosts.

The great majority of teleosts lose this open connection either at an early stage in larval development, or in post-larval life, and the adult swim-bladder is entirely closed or *physoclistous*. The swimbladder is filled either before the connection is lost, or by secretion of gas from special cells in its wall. Gas cannot be lost very rapidly from such a closed system, and this is

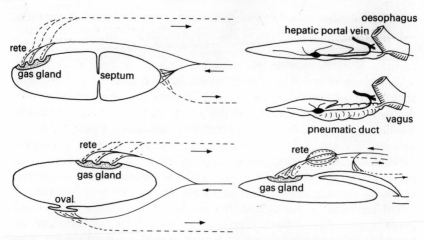

Figure 4.6 Structure of physoclistous (left) and physostomous swimbladders. Upper left: divided swimbladder (as in wrasse)—direct connection to systemic circulation on right. Lower left: gadoid swimbladder with direct connection to systemic circulation at oval. Lower right: eel swimbladder with direct connection to systemic circulation at pneumatic duct. Upper right: appearance of *Conger* swimbladder before (above) and after 5 min vagal stimulation, showing gas loss through enlarged pneumatic duct. After Denton (1961) and Fänge (1966).

why deep-water physoclists come to the surface with the viscera protruding from the mouth, blown out by the enormously dilated swimbladder which is often ruptured by the decrease in ambient pressure. As fish are not designed to be fished up, presumably the physoclistous arrangement is a better design than the physostomatous, since it is found in all advanced teleosts, though where the advantage lies is still not clear. The swimbladders of the great majority of teleosts share certain curious structural features (Figure 4.6). The connections with the blood system are complicated; there are sometimes diaphragms or sphincters isolating one part of the swimbladder from the gas inside; and parts of the lining epithelium are specially modified.

4.4.2 Gas in the swimbladder

It is known from experiments with O_2^{18} that the swimbladder gas is derived from oxygen dissolved in the surrounding water, and does not arise from chemical processes in the fish tissues; it passes into the swimbladder from the blood. So the fish using a gas-filled swimbladder has to overcome three different problems:

1. How to drive gas across the partial pressure gradient from blood to swimbladder lumen.
2. How to retain gas inside the swimbladder and prevent it diffusing out.
3. How gas may sometimes be permitted under controlled conditions to pass from the swimbladder to the blood.

a. Loss of gas. Obviously enough, the last problem is the simplest. All that is required is a capillary system to bring systemic blood into contact with the swimbladder wall. Because of the partial pressure gradient, gases will rapidly diffuse from the swimbladder into the blood, and thence across the gills into the water. If it is possible to occlude this connection with the systemic circulation at will, then the problem of controlled loss of gas is solved, even in the physoclist swimbladder. There must be a second connection with the circulation via which gases may be made to enter the swimbladder, and it is not very hard to see how this second potential 'leak' could be closed. Provided that blood passing to the system can be arranged to flow close to the blood passing away, diffusion across the walls of ingoing and outgoing capillaries will minimize the loss of gas from the swimbladder. We have already seen such a countercurrent exchange system in operation in sharks and scombroids maintaining the cruising musculature at an elevated temperature, and a similar arrangement would be effective here. The rest of the swimbladder could be made relatively impermeable if it was made of materials in which gases were insoluble, or alternatively, by making the diffusion pathway from inside to outside very long. It is not so simple to infer what structural arrangements would be required to solve the first problem. Let us see what the arrangements actually are in fishes before considering how they may operate.

In physostomes, gas can be lost from the swimbladder via the pneumatic duct, which has a valve at its junction with the oesophagus. In some fish, such as *Conger*, the duct is flattened when the fish is in buoyancy equilibrium, but if the external pressure is reduced, the duct fills with gas. Since it is supplied with a capillary plexus leading to the systemic circulation, gas can now diffuse away via these capillaries; only if the external pressure is further reduced does the oesophageal valve open, and bubbles of gas are lost via the mouth. The appropriate muscles of the pneumatic duct wall and the blood vessels are under nervous control; vagal stimulation and injection of sympatheticomimetic drugs like adrenalin cause loss of gas (Figure 4.6). In physoclists, special regions of the swimbladder wall are supplied by the systemic circulation and these can be occluded either by an adjustable diaphragm across the swimbladder (wrasse) or by a sphincter muscle closing off an oval area of the wall (gadoids and perciformes).

b. Retention of gas. The problem of retaining gas in the swimbladder is a much more difficult design problem. The swimbladder wall must be impermeable to gas, yet it is thin and elastic. Krogh (1919) found that if the swimbladder wall consisted of normal connective tissue, it would be about 100 times as permeable as it actually is. Denton and his colleagues (1972) found by ingenious experiments with *Conger* swimbladders that their strikingly low permeability resulted from the investment of the swimbladder by a layer of cells containing sheets of guanine crystals some 3 μm thick. The silvery guanine layer could be removed without damaging the underlying layers, when it was found that permeability rose forty-fold. In deep-sea fish, there is more guanine in the swimbladder wall than in *Conger*, and the impermeability results from the long tortuous diffusion pathways around the sheets of overlapping guanine crystals. We must admire the ingenuity with which fish make their swimbladders impermeable in this way, whilst still allowing them to be elastic and to change in volume, using the same material that it employs elsewhere for an entirely different purpose (Chapter 6).

c. Gas secretion. Gas enters the swimbladder via blood capillaries which run to a specially modified area of the inner wall, the gas gland. It is evident that this is where gas is secreted into the bladder, for in actively secreting bladders, the cells of the gas gland are covered with foamy mucus. The connection with the systemic circulation is a potential leak from the swimbladder, and it is here that the countercurrent capillary arrangement of the rete mirabile occurs. Anatomically, there are three sorts of retial arrangement; functionally they are the same. Either incoming and outgoing capillaries are apposed some way from the gas gland; or they may be in connection with the gas gland itself; or in some fish (salmonids) the retial system is diffuse, forming a so-called micro-rete.

In transverse section (see Figure 4.10), the enormous extent and close apposition of the retial vessels is clear, and we can well understand how the system operates to prevent loss of gases from the swimbladder. The deeper a fish normally lives, the greater the partial pressure difference between the swimbladder gas and the gases in the surrounding seawater, and in the systemic circulation. It is not surprising that the length and complexity of the retial system increases as does this partial pressure gradient, and that the longest retia yet observed (25 mm long) were found in the abyssal *Bassozetus taenia* (caught between 4575 and 5610 m).

The retia mirabilia, then, operate to reduce the loss of gas from swimbladders which are in equilibrium, and where gas secretion is not taking place. The way they work when the fish has descended in the water and requires actively to secrete gas long remained a most challenging mystery, and is still not yet completely understood. What is needed, evidently, is some mechanism which reduces the amount of gas in the venous capillaries leaving the rete for the heart, as compared with that in the arterial capillaries entering the rete. This mechanism must be a very ingenious one, for if the partial pressures of the gases in the venous capillaries of the rete are lower than those in the arterial, we should expect gas to diffuse across from the arterial to the venous capillaries and the whole secretory process would grind (or rather, diffuse) to a halt. The baffling problem seemed that the arrangement of the rete mirabilia was well adapted to *retain* gas within the swimbladder, but also to *prevent* any gas entering it! What is needed is some change in blood properties as it passes through the gas gland which will raise the partial pressure, although the actual gas *content* of the blood is decreased, so that gas will pass into the swimbladder, and at the same time, diffuse from venous to arterial capillaries within the rete.

In most fishes, we need only seek for some change in the blood which will account for a higher partial pressure of oxygen (pO_2) in the venous capillaries, for the accumulation of gases like argon and nitrogen can be explained by purely physical effects linked to the secretion of molecular oxygen. That the secretion of oxygen is the important process is shown by the facts that newly secreted gas contains a much higher % of oxygen than does gas from equilibrated swimbladders, and that in deep-sea fishes, the swimbladder gas is nearly pure oxygen (over 80%).

Elegant experiments on shallow-water fish, in particular upon eels (which have bipolar retia suitable for cannulation and investigation of blood parameters during secretion), have shown that in these fish the important change in the blood as it flows through the gas gland is *increase in acidity*. The cells of the gas gland are rich in glycogen, carbonic anhydrase, and lactate dehydrogenase, and produce sufficient lactic acid to lower the pH of the blood passing through the gland by nearly one pH unit. Figure 4.7 illustrates the structure and assumed metabolic activities of a gas gland cell. The red cell haemoglobin unloads oxygen when it is acidified, and this Root effect (Figure 4.8) is a very marked one: a change of one pH unit unloading almost 50% of the oxygen, even against considerable oxygen pressures. As arterial blood flows through the rete, therefore, it receives oxygen and acid from the venous capillaries, and pO_2 is thereby increased so that it exceeds that within the swimbladder, and gas diffuses across the wall at the gas gland into the swimbladder. The venous

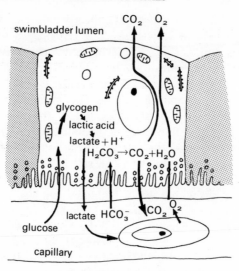

Figure 4.7 Operation of gas gland cell. After Fänge (1966).

blood leaving the gas gland has a higher pO_2 than the blood entering it, but a lower oxygen concentration (less oxyhaemoglobin) and it is also very acid. As it flows into the rete, oxygen, lactic acid and CO_2 will diffuse across into the arterial capillaries, the venous blood will become less acid, and this increase in venous pH will load oxygen on to the haemoglobin, thus decreasing pO_2. But Berg and Steen (1968) found (by using the ingenious experimental set-up of Figure 4.9) that the decrease of pO_2 following increase in pH was relatively slow, and took place outside the rete as the blood flowed back away from the rete towards the heart. The

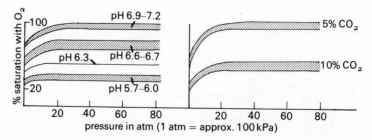

Figure 4.8 Oxygen dissociation curves for blood of black grouper (*Epinephalus mystacinus*) at 10°C, showing large Root effect even at high O_2 pressures when blood is acidified either by adding lactic acid (left) or CO_2 (right). After Scholander and van Dam (1954).

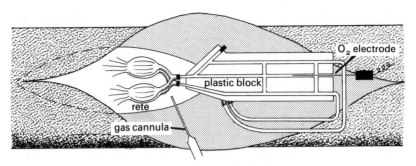

Figure 4.9 Experimental arrangement to determine changes in blood properties during gas secretion by the eel *Anguilla*. After Berg and Steen (1968).

half-time of this Root-on shift, loading haemoglobin with oxygen as pH increases, was in the eel some 10–20 seconds, compared with the Root-off unloading shift, which had a half-time of only 50 ms. Figure 4.10 shows the process of secretion in the eel in schematic form.

It is not yet clear that this scheme applies to *all* fish. Some store not oxygen, but almost pure nitrogen, and there is not as clear a correlation as we should expect between the magnitude of the Root effect and the depth

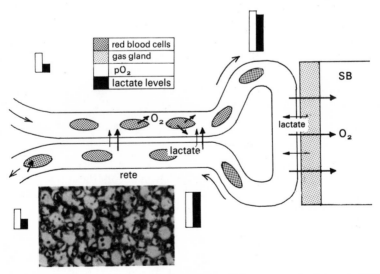

Figure 4.10 Diagram showing how the rete operates during gas secretion into swimbladder (SB). Inset shows part of the rete of *Conger* in transverse section (arterial capillaries smaller than venous).

at which a fish normally lives. One possibility is that osmotic changes in the blood within the gas gland alter the solubilities of gases and that gas is secreted in part (or in some fishes wholly) by a salting-out effect; so far, there is no direct evidence for this though it is the simplest mechanism which would explain nitrogen secretion in some salmonids.

The gradual unravelling of the mechanism of gas secretion in teleosts has occupied physiologists of the highest calibre, and is a fascinating story which we have only been able to touch on here. Denton (1961) gives a good historical account; but there are several aspects which are not yet understood and the subject is still being actively investigated.

d. Rates of gas secretion. One difficulty about the use of gas for static lift, which we have not considered, is the question of the *time* required for gas to be secreted or absorbed. Many fish, some of them with gas-filled swimbladders, make extensive vertical migrations. Are they able to main-tain constant swimbladder volume during these migrations, or must they accept loss of neutral buoyancy as they descend? Most fish can refill their swimbladders if these are artificially emptied in 4–48 h at atmospheric pressure, although some with poorly developed retia (like salmonids) may require nearly a fortnight to do so. Some myctophids descend from the surface to 400 m, and unless they secrete gas as they descend, the volume of the swimbladder will decrease (there is thus a positive feedback on the rate of descent) until at 400 m the swimbladder volume will only be 1/40th of that at the surface, and the fish will be denser than the ambient water. Several lines of approach suggest that such a fish could not secrete gas sufficiently rapidly to make itself neutrally buoyant both at the surface and at depth. First, secretion implies oxygen uptake from the water; so much oxygen would have to be secreted into the swimbladder that impossible rates of uptake at the gills are required. Secondly, making plausible assumptions about the efficiency of the secretion process, and the oxygen consumption of the fish, Kanwisher and Ebeling (1957) have calculated that a myctophid would require 33 h to descend from the surface to 400 m if it was to remain in neutral buoyancy by gas secretion as it descended. Since the fish makes this journey *daily* it seems that it cannot be neutrally buoyant at depth even if it is at the surface (see also Alexander, 1972).

We have seen in this section that there are inherent difficulties in gas storage as a solution to the lift problem, and that even if a fish lives more or less at constant depth (as at the sea bed) remarkable adaptations are needed. In the final section, we shall consider a compromise solution; fish which use both gas and lipid for static lift.

4.5 A combination of gas and lipid for static lift

Herring are primitive fish yet in some ways their solution to the lift problem is simpler and more effective than that of more advanced fish (Brawn, 1962). The swimbladder is quite different to those we have so far considered. There is no rete, and gas cannot be secreted; all the gas in the swimbladder is taken in at the surface via the open pneumatic duct. There is an open connection with the exterior from the posterior pole of the swimbladder (see Figure 9.4). At the surface, the volume of the swimbladder varies, but it is usually below the 5% of body volume required by ordinary marine teleosts. The tissues contain a good deal of lipid (up to 20%) and the fish is neutrally buoyant, swallowing less gas the more lipid there is in the tissues. As herring descend, swimbladder volume decreases, but lipid still contributes static lift, so that less dynamic lift has to be generated than if gas alone provided lift. Similarly, as the herring ascends in the water it can rise to the surface without pausing to equilibrate. Physoclist predators of the herring (like the cod) can only cope with changes in swimbladder volume of some 20%, which means that a cod adjusted for neutral buoyancy at 50 m could only rise to 37 m before absorbing gas. So the herring, able to rise directly from any depth to the surface, has a positive advantage over those predators who rely on gas alone for static lift. Of course, this is no help when herring are hunted by sharks such as the mako (*Isurus*) or blue (*Prionace*) which use lipid for static lift!

The simple arrangement in the herring is a suitable one for fish which have a wide vertical range, and although the herring swimbladder is highly specialized for sound reception (Chapter 9) it shows us today a primitive stage in the development of the swimbladder which is still a successful solution for fish which migrate diurnally. Probably many of the myctophids operate in the same way, although they need a rete since they have physoclistous swimbladders.

CHAPTER FIVE

GAS EXCHANGE AND THE CIRCULATORY SYSTEM

The mechanics and physiology of respiration are especially interesting in fishes, for there is considerable diversity of approach to the problem of acquiring and transporting oxygen, not only in fishes of different habitat, but also in fishes of different groups living in the same habitat. A much wider scope for activity is given to a scombroid, for example, than to a lamprey or a hagfish by the design of their respiratory and circulatory systems. A surprising number of fishes are facultative or obligate air-breathers, rising to the surface to gulp air into various kinds of respiratory chambers. Several of these, like the walking catfish (*Clarias batrachus*) of the southern states of America, synchronize these trips to the surface so that the water boils with a mass of fishes for a moment or two, and remains quiet until the next excursion (a strategy which apparently confuses predators). Some, like mudskippers, spend much of their lives out of water, but the great majority of fishes spend all their lives in a dense viscous fluid which contains relatively little oxygen. At the sea surface, for instance, air-saturated water contains (at 20°C), only around 3 % of the oxygen in the same volume of air.

Since oxygen only crosses cell surfaces by diffusion, this means that the gas exchangers for acquiring it must have a large area, and a relatively large expenditure of energy is needed to force water to flow around them. Alternatively, the fish has to accept a very low rate of oxygen uptake. What is more, and we need to bear this in mind when considering the design of these gas exchangers, the blood is almost always quite different in composition and osmolarity to the external water. The blood has to be in intimate contact with the water to make the diffusion pathway short; so there must be special arrangements to circumvent or at least cope with water and ion flow across the exchanger.

5.1 Origin of respiratory gills

In amphioxus and tunicates, the gills are ciliated food-collecting sieves, and blood passing through them is probably *de-oxygenated* since the ciliary tracts of the gill bars must use more oxygen than is provided by water flowing through. The respiratory gills of larger and more complex chordates presumably arose in phylogeny from a filtering system of the amphioxus type, the significant step probably being the change from ciliary to muscular movement of water through the gill. This change, leading to the arrangement seen in lamprey ammocoete larvae, may have come about as a consequence of the demand for a higher filtration rate than cilia alone could provide; only secondarily were the gills specialized as gas exchangers, when this more efficient filtering system allowed increase in size beyond that where simple diffusion across epithelial surfaces sufficed for respiration. In lampreys and many teleosts, however, a significant proportion of the oxygen needed is still gained across the skin, despite the development of respiratory gills.

5.1.1 Lampreys

In the ammocoete larva (Figure 5.1) water is driven over the gills partly by the action of an anterior muscular velum, and partly by gill muscles; in the

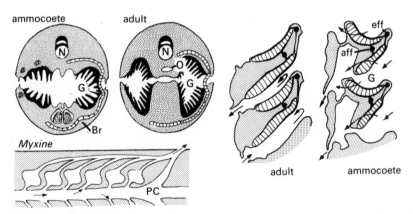

Figure 5.1 Arrangement of gills in agnathan fishes. Upper left: transverse sections of branchial region of ammocoete and adult lamprey. O: oesophagus; G: gill sac; Br: branchial skeleton; N: notochord. Right: horizontal sections of left gill sac of ammocoete and adult (anterior to top) showing tidal water flow in adult gill sac. Efferent (eff) and afferent (aff) vessels of the gill sacs are arranged to allow countercurrent flow in larva. Bottom left: horizontal section of right gill sacs of *Myxine* (anterior to left) showing pharyngo-cutaneous duct (PC) and common outflow. After Alcock (1893), Sterba (1953) and Goodrich (1909).

adult lamprey only by the action of the intrinsic gill musculature. As adults, of course, lampreys are either ectoparasites on other fishes, or like the brook lamprey (*Lampetra planeri*) they do not feed. So the lamprey gill is the simplest design we have which functions solely to produce a flow of water over a gas exchanger. Regrettably, rather little is known of respiratory physiology in lampreys.

The gills (Figure 5.1) are large muscular sacs, supported by a cartilaginous elastic branchial skeleton which lies entirely *outside* the gill sacs, quite unlike the gnathostome arrangement. Adult lampreys attach themselves to rocks (which they can move around to make nests, Chapter 8), and to the hosts on which they feed and hitch lifts (most basking sharks carry one or more sea lampreys), so that the mouth is not open: water flow over the gill lamellae is tidal, flowing into and out of the gill sacs via the valved branchial openings. In the ammocoete, however, water enters the mouth and flows out through the branchial openings, so that water flow over the gill filaments is unidirectional.

We can easily see why unidirectional water flow over the gas exchanger is a good plan, for it allows the possibility of countercurrent blood and water flow in the gas exchanger. If blood flow through the gas exchanger is in the same direction as the water flow, then the maximum partial pressure of oxygen in the blood of the exchanger would be the same as that in the exhalent water stream. On the other hand, if the blood flows in the opposite direction to that of the water, the maximum pO_2 of the blood leaving the exchanger could be very close to that of the incurrent (ambient) water, and *above* the pO_2 of the exhaled water. We have seen a similar countercurrent system (between two kinds of capillaries) in the rete of the swimbladder. In higher fishes, the result of countercurrent water and blood flow in the gills is that up to 90% of the oxygen in the water flowing into the gill chambers can be extracted from it, a remarkable degree of efficiency. No wonder that this is the arrangement found in all other fishes; presumably adult lampreys have only abandoned it in consequence of the development of their special suctorial system. In the ammocoete gill, the possibility of a countercurrent flow exists although this has not been experimentally demonstrated. In any event, this can hardly be of great efficiency, since cutaneous respiration seems to be important in both ammocoetes and adults. In brook lamprey, up to 18% of total oxygen uptake takes place across the skin, and in both larvae and adults, the gill surface area is relatively small—about twice the surface area of the skin in adults (in higher teleosts, the gills can be up to 60 times the surface area of the body.)

In line with this, oxygen *consumption* by lampreys is low (large ammocoetes of *Ichthyomyzon* consumed $0.05 \, \text{mg} \, O_2 \, g^{-1} \, h^{-1}$ when burrowed; adult *L. fluviatilis* up to $0.29 \, \text{mg} \, O_2 \, g^{-1} \, h^{-1}$), and in *Entosphenus* adults, oxygen extraction from the water was only between 10.0 and 25%. Respiratory efficiency in lampreys is therefore rather low, as we can see from another point of view if we consider respiratory efficiency as the ratio between the amount of oxygen acquired at the gas exchanger which is available for general metabolic purposes, and that used by the respiratory muscles themselves. On this criterion, efficiency seems likely to be low, for lampreys have a relatively large amount of branchial musculature.

5.1.2 Hagfishes

In the other living agnathans, the myxinoids, there are muscular gills like those in lampreys, which may open separately to the exterior, but most is known about the European hagfish, *Myxine*, where they open into internal ducts leading to a single external common aperture (Figure 5.1). Water enters via the nostril, passes along the nasal duct, and thence to the velar chamber via a nasopharyngeal duct. A complicated set of velar muscles and cartilages are used to pump water backwards either into the gill chambers, or, by-passing them, to the exterior via the pharyngo-cutaneous duct.

This complex system results in water flow that is always unidirectional over the gill lamellae; unlike lampreys, ventilation is not tidal. However, since hagfish can survive well if their nostrils are blocked (thus interrupting gill ventilation), there is little doubt that cutaneous respiration is even more important than it is in lampreys. This is scarcely surprising when we look at the vast sub-cutaneous blood sinuses that are found in hagfishes: if one holds up a living hagfish, blood in these sinuses flows downwards and swells the lower part of the body, and indeed, *Myxine* is rather like a lamprey that is surrounded by a layer of blood contained in a thin outer skin. Blood volume in hagfishes is about double that of lampreys, and lampreys themselves have an unusually high blood volume compared with other fishes. Although oxygen consumption of lampreys is apparently low, they have relatively large hearts as compared with other fishes: this puzzling observation remains to be further investigated.

Much more is known about respiration in gnathostome fishes, particularly in teleosts, and in the remainder of this chapter we shall consider the design of their respiratory and circulatory systems.

5.2 Gnathostome fishes

Almost all studies have been on advanced teleosts (acanthopterygians) but there is no reason to suppose that gill design has changed much in teleost phylogeny, for eel gills are like those of acanthopterygians, and eels are fairly close to the primitive teleost stem (Figure 1.3). Figure 5.2 shows the basic design of a teleost gill. Four gill arches bear regular comb-like gill filaments, and on each there are closely-ranged primary gill lamellae. Each primary lamella bears a number of stacks of smaller secondary lamellae which are parallel to the water flow through the branchial chamber: it is these secondary lamellae that are the actual site of gas exchange. Their numbers and dimensions vary between different species, but in fish weighing around 1000 g there may be up to 18 000 cm^2 of secondary lamellae. This very large surface is needed partly because the oxygen content of water is low, and partly because rates of oxygen diffusion are relatively low in animal tissues. (In connective tissue, for example, the oxygen diffusion rate is only 10^{-6} of that in air.)

The secondary lamellae are essentially thin-walled sacs through which blood circulates, and over which water is passed. The lumen of the sacs is

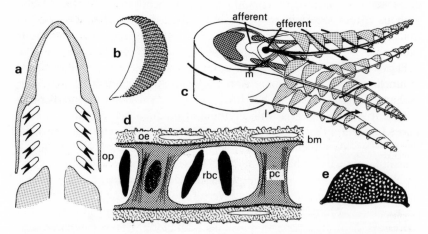

Figure 5.2 Structural design of teleost gill. (*a*) Horizontal section showing disposition of gill filaments (black) on gill arches. (*b*) Single hemibranch. (*c*) Arrangement of secondary lamellae (l) and water flow over them. Heavy stipple, gill rays and gill arch; light stipple, secondary lamellae. Efferent (eff) and afferent (aff) vessels pass to either side of the filament. Intrinsic muscles (m) can change the apposition of gill filaments. (*d*) Section across secondary lamella showing pillar cells (pc), red cells (rbc) and outer epithelium (oe) with lymph spaces and underlying basement membrane. (*e*) Cast of vascular spaces in secondary lamella. After Munshi and Singh (1968) and Hughes and Grimstone (1965).

maintained by a series of regularly-spaced bridges from one epithelial layer to the next; these bridges are formed by pillar cells that make a series of posts around which the blood flows in the lamella. Perhaps it is best to think of the blood in the secondary lamellae as flowing through an interrupted thin-walled sinus, rather than through a series of capillaries. The inner wall or lining of this sinus is formed by flanges from the bases of the pillar cells, so that to gain access to the blood, oxygen has to pass from the water across an epithelial layer; the basement membrane of the epithelial cells; and the inner endothelial (pillar cell flange) lining of the sinus. This diffusion barrier differs in thickness in different fish (as we might have guessed, it is thinnest in those fish requiring the greatest rate of oxygen uptake—Table 5.1). Apart from the specialized pillar cells, the arrangement is rather like that seen in lungfish or tetrapod lungs, but there are complications which arise from the intimate proximity of the blood to the water flowing around the secondary lamellae.

The large area of the gas exchanger, and the difference in composition between the blood and the water provides the fish with a structure that will act not only as a gas exchanger, but also as an efficient heat- and ion-exchanger. Oxygen demand varies depending on whether the fish is at rest, or is actively swimming. Thermal diffusion is much more rapid than gaseous diffusion, and fish can only maintain a body temperature above that of the ambient water by organizing special countercurrent heat exchangers in the blood system near the organs which are to be kept warm (Chapter 2). But it is easy to see that defensive adaptations are possible to minimize ion-exchange, provided that blood flow through the gas exchanger can be varied according to oxygen demand, or, alternatively,

Table 5.1 Diffusion distances between water and blood in the secondary lamellae of different fishes. Distances in μm. Note minimum distances in the active pelagic skipjack and mackerel, as compared with the other benthic fishes. From Hughes and Morgan (1973).

Fish	Epithelium	Basement membrane	Pillar cell flange	Total water–blood (mean)
Dogfish				
(S. canicula)	2.38–18.48	0.3–0.95	0.37–0.71	11.27
Squalus	3.0–22.5	0.3–0.6	0.12–0.6	10.14
Raja clavata	0.5–11.5	0.13–0.63	0.03–1.13	5.99
Microstomus kitt	0.21–16.7	0.1–0.69	0.1–0.13	3.23
Skipjack				
(Katsuwonus)	0.013–0.625	0.075–1.875	0.017–0.375	0.598
Mackerel				
(S. scombrus)	0.165–1.875	0.066–1.0	0.033–1.75	1.215

that ventilation of the gas exchanger can be varied. There has been much dispute about the first possibility, since Steen and Kruyse (1964) first provided evidence which suggested blood could pass from afferent to efferent arteries within the gill filaments without passing across the lamellar gas exchanger. This concept of a double circulation in the gill filaments, allowing blood to be shunted into a 'non-respiratory route' when oxygen demands are low, and passed through the lamellae (the respiratory route) during exercise when oxygen demand rises, has on the whole turned out to be unacceptable, although it seems such a sensible idea for the fish! Careful examination of gill filaments with latex injection has not shown any shunts which could allow red blood corpuscles to bypass the secondary lamellae, and although recent ultrastructural studies (Vogel et al., 1976) have definitely shown the existence of direct links between both afferent and efferent arteries, and the central blood space of the filament, these seem to be too small to permit red cells to pass. What is more, the relative pressures in the central space and the arteries do not suggest that flow of plasma can be significant. Yet although the present evidence does not support a dual filamentar circulation, this does not rule out the possibility that sphincters of the kind found at the bases of the arteries of the gill filaments, or at the entry to the lamellae in dogfish could alter blood flow through the gills. Circulating catecholamines (predominantly adrenalin) reduce branchial vascular resistance. There is no doubt that the resistance to blood flow in the gills can be altered (perhaps by vagal control of such sphincters), and that changes in arterial pO_2 can be produced by injection of adrenalin (increase), and acetylcholine (decrease). What is certain is that not all secondary lamellae are maximally ventilated except during exercise, and probably this results not only from changes in blood flow patterns (perhaps even in the lamellae themselves, if pillar cells are really contractile, as their morphology suggests), but also from changes in water flow within the gill.

The angle at which the two rows of filaments on each branchial arch lie with respect to each other can be changed by the operation of the intrinsic musculature of the gill filaments, thus altering the ventilation of the lamellae, but how significant such changes in gill geometry may be in normal fish is not known. In elasmobranchs (with plate gills), the filaments cannot be moved in this way, as they are fused along most of their length.

Several kinds of evidence strongly suggest that during exercise lamellar recruitment results in an increase in functional gill area, which not only increases oxygen uptake, but also necessarily increases water and ion fluxes across the gills. Farmer and Beamish (1969) found in Tilapia that plasma osmolality increased after exercise in sea water, and decreased after exercise in fresh water. Similarly, rainbow trout in fresh water lost sodium by diffusion across the gills during exercise, and gained water to such an extent that the body weight increased until urine production exceeded gill water influx.

It is hardly surprising that the cost of osmoregulation in fishes is high, perhaps around 15% of the total oxygen uptake (Farmer and Beamish, 1969). Apart from the role of the kidney in ion and water regulation, there are mechanisms on the gills themselves which regulate water and ion fluxes between water and blood. We shall consider osmoregulation and ion-exchange in Chapter 6, but it is important to realize that fish gills are unavoidably important sites of ion and water exchange, and that a significant part of the energy budget of the fish must be devoted to coping with this problem of exchange. Perhaps, as suggested in Chapter 2, use of lactate by the gill oxidative systems to provide the ATP for this requirement may be a helpful way of lowering blood lactate produced by the anaerobic activity of the white myotomal musculature.

5.2.1 Branchial pumps

Let us now consider how water flows over the respiratory surfaces of the secondary lamellae. In agnathan fishes, velar and intrinsic gill muscles are involved in operating a pump; in gnathostome fishes, similar pumps are found, although they are not necessarily always in use, since when swimming forwards above a certain speed many fishes simply ventilate the gills by opening their mouths, and no longer require to pump water through the system using branchial muscles.

In both teleosts and elasmobranchs, a double pump is used to drive water across the gill chamber: a pressure pump lying anterior to the gill resistance, and a suction pump posterior to it. It is difficult to see how these pumps operate directly, but by a combination of pressure records taken via cannulae placed at different points, strain gauge records of movements of mouth and gill openings, and simple wire electrodes within different muscles to record electromyograms, a fairly complete picture of the way the pumps operate can be obtained. With these techniques, Hughes and Ballantijn examined various teleosts and elasmobranchs; their work on dogfish (1965) is an example of the approach.

The skeletal structures involved, and the main muscles attached to them are seen in Figure 5.3; Figure 5.4 shows the relationship of the pressures observed in the orobranchial cavity and parabranchial cavities during

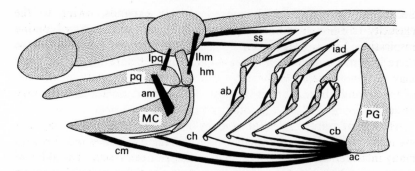

Figure 5.3 Skeletal structures and muscles involved in branchial pumps of dogfish (*Scyliorhinus*). MC, Meckel's cartilage; ab, adductor branchialis; ac, arcualis communis; am, adductor mandibulae; ch, coraco-hyoideus; cm, coraco-mandibularis; hm, hyomandibula; iad, interarcualis dorsalis; lhm, levator hyomandibulae; lpq, levator palatoquadrati; PG, pectoral girdle; pq, palatoquadrate. After Hughes and Ballantijn (1965).

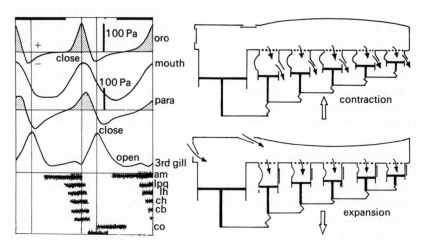

Figure 5.4 Dogfish respiratory pumps. Left: orobranchial (oro) and parabranchial (para) pressures (stippled above ambient) and movements of mouth and 3rd gill slit. Below, EMG activity from different muscles. Right: model of system in expiratory (upper) and intake phases. After Hughes and Ballantijn (1965).

respiratory movements, when muscular activity is monitored electromyographically. There are three main phases in the cycle. First, when most of the respiratory muscles are active, the volume of the orobranchial chamber is reduced, and after an initial increase (due to water flowing out of the orobranchial cavity) the parabranchial cavities also decrease in size.

Secondly, the orobranchial cavity passively expands, owing to the elasticity of the skeletal and ligamentous elements compressed during inspiration. No muscles are active during the expansion process. Lastly, there is a pause, before the cycle begins again. Before the pause, there may be a more rapid expansion of the orobranchial cavity, at the end of the second phase, during which the hypobranchial musculature may be active.

As we should expect from the anatomical arrangements shown diagrammatically in Figure 5.3 and as Hughes and Ballantijn emphasize, there are many interactions between the different parts of the system, and the two pumps are not completely separate. For example, the muscles which drive the orobranchial pump also affect the parabranchial pumps. Figure 5.4 shows a model of the dogfish system which incorporates some of these interactions. The most important feature of the system is that by using the interaction between an anterior pressure pump and a posterior suction pump, water flow over the gills is unidirectional, and so countercurrent water and blood flow are possible at the secondary lamellae. In teleost fishes, the arrangement of the dual pump is essentially the same as in the dogfish, though since the skin is stiffer in teleosts, and there is normally a stiff operculum, coupling of the two pumps is closer than in dogfish, and both expansion and contraction phases of the pumps are active. In different teleosts, the relative importance of the suction and pressure pumps differs (Figure 5.5) the suction pump being most important in bottom-living fishes.

Under different conditions (for example, lower ambient pO_2 or higher pCO_2) respiratory patterns change to cause more water to flow through the branchial chambers. This may be brought about by increasing the volume of water pumped during each cycle (increasing the stroke volume), by increasing the cycle frequency, or by a combination of the two; in

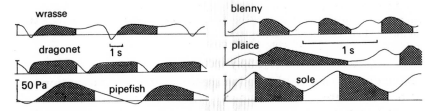

Figure 5.5 Buccal and opercular pumps in different fishes. Pressure differences between buccal and opercular cavity (+ve values: greater pressure in buccal cavity). Stippled: operation of opercular suction pump. After Hughes (1970).

different fish, each of these is seen. In trout, for example, ventilation
frequency remains much the same, even though a trout may increase the
volume of water pumped by 5 times. The regulation of the muscles driving
the respiratory flow has been studied in some teleosts, where both length
and tension receptors are involved, but these receptors (equivalent in some
respects to the familiar Golgi tendon organs and muscle spindles of
terrestrial vertebrates) have not yet been identified histologically in the
muscles.

5.2.2 Ram ventilation

Many teleosts, and probably most elasmobranchs, cease respiratory
movements when they are swimming fast enough that water flow through
the mouth suffices for respiration. Some, like the larger scombroids, are
unable to respire except by this ram-jet method. Roberts (1975) has
investigated the way in which fish in a respirometer tube ram-ventilate
above a certain swimming speed, and has found that the transition to ram
ventilation takes place when the pressure difference across the gills exceeds
2 kPa. An earlier analysis of ram ventilation in tuna by Brown and Muir
(1970) showed that the dynamic pressure required in skipjack tuna was
around 80 Pa, comparable to the pressures of 50–100 Pa developed in
resting fish of various species using the branchial pump mechanism of
passing water over the gills.

However, the cost of ram ventilation is lower than that of operating a
branchial pump. Brown and Muir estimated that 1–3 % of the total energy
expended by the skipjack was required for ram ventilation of the gills, and
(although estimates of this kind are obviously subject to errors) in slowly
swimming or resting fishes of other species using the branchial pump
system, it seems that as much as 15 % of the total energy expended is
required for ventilation. Magnuson (1978) has re-examined ram ventilation in tuna, and has revised Brown & Muir's estimate upwards to around
9 % of the total energy; still better than pumping. No wonder that many
demersal fishes change to ram ventilation as soon as their swimming speed
rises sufficiently to maintain around 100 Pa across the gills; in the special
case of the remora which lives attached by its dorsal sucker to larger fishes,
ram ventilation is obtained at no cost to the remora! Obviously, when
ram-ventilating, fishes can simply increase the mouth gape to increase
water flow over the gills at a given speed, or alternatively, may increase
swimming speed but both cause large increases in drag, and require
additional work (and oxygen consumption) from the propulsive

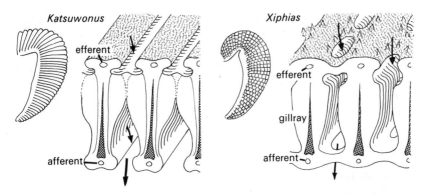

Figure 5.6 Gill filament modifications. The gill filaments of skipjack tuna (*Katsuwonus*) and swordfish (*Xiphias*). Arrows indicate direction of water flow; opposite to blood flow. Insets: hemibranchs of first gill arches. After Muir and Kendall (1968).

musculature, so that, as Roberts (1975) suggested, it is probable that the fish monitors the rate of flow of water through the gill chamber by flow receptors in the chamber, rather than by a slower pO_2 receptor system.

Tuna and swordfish gills do not look like those shown in Figure 5.2—the gill filaments are linked together by a series of bridges, and the filaments may be fused at their edges (Figure 5.6). A somewhat similar arrangement appears in the holostean *Amia*. It is difficult to think of two groups of fishes which are more unlike, either in lifestyle (*Amia* is a lurking predator in fresh waters of the northern USA) or in taxonomic position, so we have to think of two different functional explanations for this convergence. Since tuna and marlin gills often show regenerated and damaged areas, it seems likely here that fusion is an adaptation to strengthen the gas exchanger against damage by floating objects in the rapid inhalent flow. But in the sluggish air-breathing *Amia*, fusion may have arisen in order to keep the gill sieve patent in air. *Amia* is unusual in being an air-breather living in temperate regions, for most of the fascinating variety of air-breathing fishes live in tropical swamps, where a combination of stagnant water, high temperature and abundant microorganisms make the water very acid, with a high CO_2 and low O_2 content. These unfavourable conditions for aquatic respiration have led to extraordinary adaptations for acquiring oxygen, also found in fishes which live out of the water normally, or during droughts.

In the swamps, some fish manage by ventilating the gills with water obtained from just below the surface where it is oxygenated, but most have to use accessory respiratory organs of various kinds.

5.2.3 Air-breathing fishes

All fishes which breathe air contain a hollow space whose walls are richly vascularized, and which can be ventilated periodically. A very few, like the eel *Anguilla*, use the unmodified gills in the branchial chamber as the respiratory organ, but almost all have either modified existing structures or developed entirely new ones, and where aerial respiration is dominant, the gills are much reduced. The gills are not lost, even in lungfishes, the most modified of all air-breathing fishes, for they still act as the site of CO_2 excretion, together with the skin. However, as fishes in water with low ambient pO_2 face the risk of oxygen loss at the gills, these are often modified to increase the diffusion pathway. For example, in the climbing perch, *Anabas*, the diffusion distance is $15\,\mu m$ compared to $1–3\,\mu m$ in normal water-breathing fishes (Table 5.1). Since CO_2 diffuses so much more rapidly than O_2, this will have little effect on CO_2 excretion. Alternatively in facultative air-breathers, like *Amia*, or the lungfish *Neoceratodus*, O_2 loss at the gills is avoided when the pO_2 of the water falls, by shunting blood from the gill circulation to that of the swimbladder or lung.

In fishes respiring in water with gills, the gas exchanger is perfused with systemic venous blood, and sends oxygenated blood directly to the

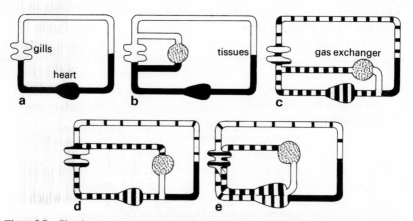

Figure 5.7 Circulatory patterns of gills and air-breathing systems organs in air-breathing fishes. (*a*) Normal fish (gills in series with systemic bed tissues). (*b*) Air-breathing organ opercular chambers or buccal mucosa (*Clarias, Saccobranchus*). (*c*) Air-breathing organs opercular or pharyngeal mucosa (*Electrophorus, Anabas, Periophthalmus*). (*d*) Air-breathing swimbladder (holosteans). (*e*) Swimbladder lung-like, partial division between pulmonary and branchial circulations (lungfishes). Oxygen content of blood shown by hatching: black, low; white, high. After Johansen (1974).

systemic arterial system, as in Figure 5.7. In some air-breathers, accessory respiratory organs derived from the gills, buccal cavity and opercular chambers are in parallel with the gills, and their blood supply is derived from that of the gills in such a way that blood passes through them and then joins oxygenated efferent blood from the gills, passing to the systemic arterial circulation. But other air-breathers do not have this efficient arrangement, and the oxygenated blood from their gas exchangers passes to the systemic venous circulation anterior to the heart. When the gas exchanger is the swimbladder or the lung (in lungfish) special afferent vessels pass to the gas exchanger from the branchial circulation. In lungfish, the gas exchanger drains directly to the heart by a pulmonary vein, just as in terrestrial animals (Figure 5.7).

The gas exchangers themselves are very varied, sometimes being diverticula of the buccal cavity (as in the catfish *Clarias*), or specially vascularized parts of the intestine, which the fish ventilates by gulping air, and burping it out of the mouth or through the anus. In the electric eel *Electrophorus* (Figure 5.8) the whole buccal mucosa is papillated and highly vascularized, making a respiratory surface around 15 % of the total surface of the fish. *Electrophorus* is an obligate air-breather, and although it seems remarkable that a delicate respiratory surface in the mouth can operate in a fish that feeds on living prey, this is probably because *Electrophorus* stuns its prey with a powerful electric shock (Chapter 9) and does not bite or chew its food.

Very few fishes use the oesophagus as an accessory respiratory organ, for the same problem applies: the Alaskan blackfish (*Dallia*), which feeds on

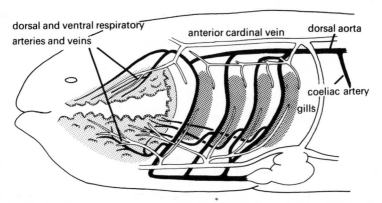

Figure 5.8   Respiratory buccal mucosa and its blood supply in the electric eel *Electrophorus*. After Johansen *et al.* (1968).

small crustaceans, lives in sub-arctic swamps which become very low in pO_2 during the summer; like the tropical *Menopterus* it respires via the oesophagus.

Lungfishes are especially interesting, because they resemble amphibians in the way that they obtain oxygen. *Neoceratodus*, the Australian lungfish, is the most primitive, and normally breathes water; it has only one lung. The south American *Lepidosiren* and the African *Protopterus*, are obligate air-breathers, and have paired lungs (Figure 5.7). The latter have a complicated septated lung divided up finally into alveolar pockets very similar to those in terrestrial vertebrates, and more complex than the simple lung of many urodeles. Unlike most air-breathing fishes, lungfish can use their lungs and gills simultaneously—just as in many amphibians there is concurrent pulmonary and cutaneous gas exchange. To do this efficiently, lungfish can adjust their circulatory system so as to favour gas exchange by one or other route. This was shown by Johansen and his colleagues (1968) in a series of ingenious experiments where they sampled blood from different sites in the circulatory system, and adjusted the pO_2 of the water the fish was living in. They measured the pO_2 of the blood at the different sites shown in Figure 5.9 and converted these values to oxygen content (to account for the oxygen combined with haemoglobin) using O_2-equilibrium curves. Obviously, the oxygen content at different points in the system allowed estimates of the degree to which blood was selectively passed through different routes, and of the relative importance of the two gas exchangers. Table 5.2 shows results with *Neoceratodus* and *Protopterus*.

In well-oxygenated water, *Neoceratodus* does not ventilate the lungs, which therefore have no respiratory function. Nevertheless, rather surprisingly, the fish sends about the same amount of blood to the heart from

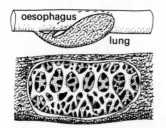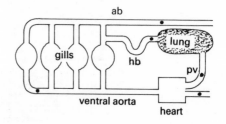

Figure 5.9   Left: lung of *Protopterus* showing septated structure. Right: sampling sites in experiments of Johansen *et al.* (see Table 5.2). After Spencer (1893). pv, pulmonary vein; ab, anterior branchial; hb, hemibranch.

Table 5.2 Blood O_2 content in two lungfishes under different conditions. (From Johansen et al., 1968)

| Species | Condition | O_2 content (vol. %) | | | | | Pulmonary venous blood/ vena cava blood | |
		Pulmonary artery	Pulmonary vein	Anterior branchial	Vena cava	Anterior branchial	Pulmonary artery
Neoceratodus	In aerated water	7.3	7.25	5.0	3.4	5/4	
	In hypoxic water	6.0	7.9	6.75	0.8	5/1	3/1
Protopterus	In aerated water	4.3	6.05	5.5	0.15	10/1	7/3

the pulmonary vein as from the vena cava. But when the pO_2 of the ambient water is lowered to 5.3–10.6 kPa *Neoceratodus* begins to breathe air, and now the measurements of Table 5.2 show that although some oxygenation of the blood takes place at the gills, by far the most important site of gas exchange is the lung, where the pO_2 of the blood rises from 3.3 kPa to 12.6 kPa as it crosses the lung. Blood in the anterior branchial arteries (which supply the systemic circulation) is now made up of about 5 parts of pulmonary vein blood and 1 part of vena cava blood. Similarly, blood passing to the lung in the pulmonary artery is made up of about 2 parts of pulmonary venous blood and 1 part of vena cava blood. So *Neoceratodus* can partially separate the blood leaving the heart into streams flowing to the anterior and to the posterior branchial arches, and as we should suspect (if we did not already know) the heart is partially divided and there is a rudimentary spiral valve in the sinus and conus, foreshadowing that of amphibians. In *Protopterus* more efficient separation of pulmonary vein and vena cava blood is found, and blood in the anterior branchial artery contains about 10 parts of pulmonary vein blood to one of vena cava blood.

Although lungfish have 'internal' nares (Chapter 1), they breathe through the mouth, taking a gulp of air into the expanded buccal cavity which has previously been emptied of water, and then deflating the lung by intrinsic muscles, before closing their mouths and opercula and forcing air into the lungs by raising the floor of the buccal cavity. In water, *Protopterus* breathe every 5–7 minutes, but if kept out of water (which they support perfectly well) they breathe more rapidly, every 1–3 minutes.

5.2.4 Aestivation

Air-breathing in lungfish enables them to live in water of low oxygen content, and to flounder across from a dried pool to another where water remains, but both *Lepidosiren* and *Protopterus* (not *Neoceratodus*) can survive a prolonged dry season by another remarkable habit: they aestivate. As the water dries away and the pool becomes more and more muddy, the fish burrow into the mud and become torpid, reducing their metabolic rate and oxygen demand until reawakened by the first rains. *Protopterus* makes a bottle-shaped burrow lined by the mucus from the skin, thus forming a cocoon; the nares are plugged with mucus and the fish breathes air through the mouth from the tube leading to the surface once an hour or so. In nature, aestivation lasts some 4–6 months, but in laboratory experiments, aestivating fish have survived to be reawakened after several years. When water enters the tube to the surface, and reaches the mouth, the fish makes breathing movements, and after a series of convulsive jerks, awakens to swim out of the burrow. The habit of aestivation is an ancient one, for lungfish burrows have been found from the Devonian and Carboniferous and the habit has been continued in the amphibians.

5.2.5 The circulatory system

So far, we have looked at ways in which fishes gain oxygen from the external environment, whether water or air, and have examined the structure of the gas exchangers. Having acquired oxygen and lost carbon dioxide at these gas exchangers, how do they transport these gases around the body?

With the exception of lungfishes, which, as we have seen, are rather like amphibians in the design of their circulatory system, all fishes have a single circulation with a simple heart with four chambers: sinus venosus, atrium, ventricle, and either a bulbus or a conus, leading to the ventral aorta. In elasmobranchs, the conus arteriosus is contractile, and contracts in sequence with the rest of the heart, but in teleosts, the equivalent bulbus arteriosus is simply an elastic reservoir which is passively enlarged by blood driven forwards from the ventricle (Figure 5.10). Valves at the junctions between the different regions (and in a series along the conus) assure

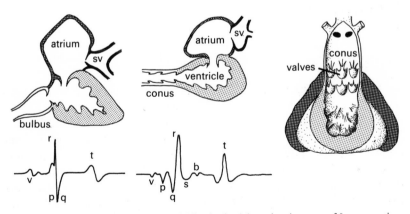

Figure 5.10 Fish hearts. Left, teleost; middle, shark; right, valves in conus of *Isurus*. sv, sinus venosus. Lower left, electrocardiograms of trout; lower right, electrocardiogram of shark (*Heterodontus*). v, pqrst and b: deflections associated with depolarizations and repolarizations of different regions of the heart (see text). After Bennion (1968), Daniel (1922) and Holst (1969).

unidirectional flow through the system. The holosteans *Amia* and *Lepisosteus*, and the chondrostean *Polypterus*, resemble sharks in having a rigid pericardium and a contractile conus—in *Lepisosteus* there are no fewer than 72 valves arranged in 8 rows along the conus.

The different regions of the heart contract sequentially, and so, just as in ourselves, it is easy to record an electrocardiogram which corresponds to the sum of the electrical events during the cycle. In teleosts, the ECG is rather like that in mammals, but in elasmobranchs additional deflections (the V and B waves, resulting from depolarization of the sinus venosus and of the conus) are seen (Figure 5.10). As yet very little is known of the ways in which the cardiac impulse is delayed along the contraction route, nor of the location of the pacemakers in most fishes, although in some sharks (including *Heterodontus*), and in *Anguilla*, the heart pacemaker seems to lie at the sinu-atrial node, so that the V-wave precedes the other events of the heart cycle. Ventricular contraction naturally leads to cyclic variations in flow and pressure of the blood leaving the heart for the gills, but these pulsations are minimized as far as possible, for constant pressure and flow are needed at the gas exchanger. In the ling-cod (Figure 5.11) Stevens and his colleagues (1972) found that at a flow rate in the ventral aorta of $10 \, ml \, min^{-1}$, the blood flow due to elastic rebound of the bulbus amounted to around 29% of cardiac output; the bulbus performed a significant role in maintaining blood flow in the aorta during ventricular relaxation (diastole). In this way, the teleost bulbus damps the pressure and flow oscillations in the ventral aorta, and in fishes with an elongate valved conus, this acts in a similar way, since contraction along the conus is slow (which is why there are so many valves in series along it).

The elasmobranch heart is unlike that of teleosts, for it is filled not

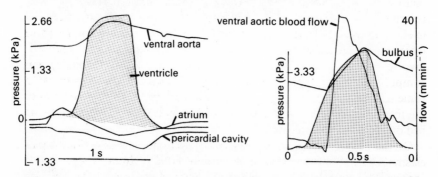

Figure 5.11 Pressures in different regions of the heart and associated vessels in a shark (*Mustelus*), left, and a teleost (*Ophidion*) right. After Sudak (1965) and Randall (1970).

simply by the *vis a tergo* of the blood returning to it (as our heart is), but blood is actually sucked into the atrium, which during diastole is below environmental pressure (Figure 5.11). The elasmobranch heart (like the dipnoan), lies in a more or less rigid, non-compliant pericardium, familiar to anyone who has dissected a dogfish, so that contraction of the heart leads to reduced pericardial and arterial pressures. Blood flows into the sinus venosus and thence to the atrium from the large posterior cardinal sinuses which provide a reservoir allowing most of the blood return to the heart to take place during the contraction of the ventricle (systole).

Regulation of the heart in all fishes except hagfishes is under vagal control, vagal stimulation causes marked slowing of the heart, probably via cholinergic fibres. The discovery that the hagfish has no nerve supply to the heart was made by Greene in 1902, who found that neither vagal stimulation, nor any other nerve stimulation could produce changes in heart rate. Luckily, he had been conscientious enough to try the experiment out before giving it to his students! More recently, evidence has been accumulating that adrenergic and purinergic fibres also affect the fish heart.

It may seem surprising that the heart in many fishes is not the only source of the pressure increase required to drive blood around the circulation. In hagfish, for example, there are caudal hearts, which have valved connections with the caudal vein, portal hearts contracting to drive blood through the common portal vein supplying the liver, and cardinal hearts bringing blood along the anterior cardinal veins. Furthermore, as Johansen and Strahan (1963) showed, the gill musculature is important in assisting flow of blood into the dorsal aorta, producing low slow pressure pulses clearly different to those resulting from ventricular contractions. In many elasmobranchs, there are accessory mechanisms for assisting venous return (we have already seen that venous return pressure is extremely low in elasmobranchs, and that the venous system has large sinuses in it); these mechanisms are essentially similar to the myxinoid caudal heart, or involve valves allowing blood to pass from parts of the venous system compressed during respiratory or body movements into the large veins. In some teleosts, as in elasmobranchs, there are arterial valves at the bases of the intercostal arteries, which act to prevent backflow from the systemic bed during swimming movements.

On the whole, the teleost circulatory system is more efficient than that of elasmobranchs; blood volume is lower and blood pressures (in particular venous return pressures) are higher, and instead of large sinuses, teleosts have narrower veins.

5.2.6 Fish blood

The blood in the circulatory system transports oxygen to the tissues, and removes CO_2 from them. It varies in its properties in different fishes depending on the metabolic demands and the way in which the fish obtains oxygen and excretes CO_2. For example, blood in scombroids must have a much higher oxygen capacity than that of sluggish fishes such as rays; in obligate air-breathers, it must be less sensitive to carbon dioxide content than it is in water-breathing fishes. Some values for the oxygen capacity of whole blood in different fishes are seen in Table 5.3. The oxygen capacity of whole blood comprises the oxygen in solution in the blood, and the oxygen combined with haemoglobin; this respiratory pigment in the blood raises the oxygen capacity up to 40 times.

Table 5.3 Blood O_2-capacity and gill areas in fish of different habit (after Steen, 1971).

Species	O_2-capacity (vol. %)	Gill area ($mm^2 g$ body $wt.^{-1}$)	Habit
Bonito (*Sarda*)	18.0	595	
Mackerel (*Scomber*)	19.6	1158	Very active
Menhaden (*Brevoortia*)	16.2	1773	
Butterfish (*Pholis*)	10.7	598	
Sea-robin (*Prionotus*)	9.3	360	Active
Eel (*Anguilla*)	8.0	302	
Goosefish (*Lophius*)	5.7	196	
Toadfish (*Opsanus*)	5.3	200	Sluggish
Sand-dab (*Hippoglossoides*)	4.6	188	

Fishes lacking haemoglobin. Remarkably enough, in several Antarctic ice-fishes of the family Chaenichthyidae (Figure 2.9), haemoglobin in the blood is much reduced or absent, and there are no red blood corpuscles. All the oxygen reaching the tissues must do so in solution in the blood, which has the same oxygen capacity as sea water, around 0.7 volume % compared with around 8 volumes % in many normal fish with haemoglobin (Table 5.3). This seems a strange adaptation (shared only by eels, which abandon haemoglobin in their leaf-like transparent lepto-cephali larvae in order to increase transparency), and no-one has yet suggested a reason for its existence in icefish. However, to overcome the low oxygen capacity of the blood, icefish have relatively large gills, relatively large hearts assuring rapid blood flow, assisted by large blood vessels, and to reduce oxygen demand, they have reduced their red musculature (Chapter 3).

Resting oxygen uptake of icefish is around $\frac{1}{2}$ to $\frac{1}{3}$ of fishes in the same habitat which possess haemoglobin, so by a combination of low metabolic rate and high circulation rate, they exist without haemoglobin. In trout, pike and goldfish, oxygen dissolved in the blood can be shown to suffice for resting metabolism, for such fishes poisoned with CO (so that the haemoglobin cannot combine with oxygen) survive well until exercise increases oxygen demand, when they perish.

Oxygen transport by haemoglobin. Apart from the special case of the icefishes, oxygen is mainly transported from the gas exchanger to the tissues by haemoglobin, oxygenated when pO_2 is high, deoxygenated when it falls, according to the oxygen dissociation curve (blood pO_2 plotted against the amount of O_2 bound to the haemoglobin). These curves are always non-linear (Figure 5.12), and usually sigmoid, which means that at the tissues oxygen can be unloaded even if the tissue is not at a low pO_2. The steep part of the dissociation curve usually represents the normal working range in the animal, so that the amount of oxygen released for relatively small changes in pO_2 is large.

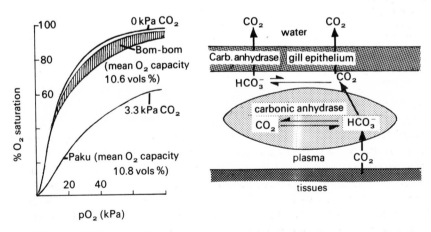

Figure 5.12   Oxygen and CO_2 transfer in fishes. Left: oxygen dissociation curves of blood from fishes living in oxygenated and oxygen-depleted water. Note difference between Paku (*Pterodoras granulosus*) which lives in well-oxygenated water and shows a large Root effect, and Bom-bom (*Myleus setiger*) which lives in oxygen-depleted water (stippled) and shows only a small Root effect. Right: scheme of CO_2 transfer in fishes. After Willmer (1934) and Haswell and Randall (1978).

We have already seen (Chapter 4) that changes in the shape of the dissociation curve can be produced by pH changes (this property being used by the fish to drive oxygen into the swimbladder from the rete). As the blood becomes more acid, the curve is usually shifted markedly to the right. This Bohr shift is found in ourselves as well as in fishes, and results from pH-dependent configurational changes in the haemoglobin molecule which inhibit oxygen binding. In practice, this means that oxygen is unloaded at sites where pCO_2 is high, just where it is needed by the animal. Mammalian blood showing this Bohr shift can still be saturated with oxygen even at increased pCO_2, but in many fishes, increase in pCO_2 not only shifts the dissociation curve to the right, but also prevents complete oxygenation of the haemoglobin, thus *depressing* the curve. Such a Root shift is really an extreme case of the Bohr shift, and effectively decreases the oxygen capacity of the blood (Figure 5.12).

Air-breathing fish like the lungfishes have haemoglobins that are relatively insensitive to pCO_2, and they need this property, because the pCO_2 at the gas exchanger will be higher than in water-breathing forms, and pCO_2 is higher in the blood (Figure 5.12). A similarly reduced Bohr shift is found in other fish which inhabit stagnant oxygen-poor waters of high pCO_2, such as the South American electric eel, *Electrophorus*.

Carbon dioxide transport. Less is known of the way in which the blood transports CO_2 from the tissue and excretes it at the gills or skin. A recent scheme proposed by Haswell and Randall (1978) is shown in Figure 5.12. CO_2 is excreted at the gill either as molecular CO_2 or as bicarbonate ion, mainly from plasma bicarbonate. It passes into the blood at the tissues, and enters the red cells, where it is rapidly rehydrated to bicarbonate (this reaction is catalyzed by carbonic anhydrase in the red cell); as a result oxygen is driven off the haemoglobin. Since water enters to rehydrate the CO_2, the red cells swell, for there is no chloride shift exchanging chloride for bicarbonate across the red cell membrane, as there is in mammals. CO_2 is also rehydrated in the plasma, but at a much lower rate than in the red cells, so the rise in plasma bicarbonate takes place after the blood has left the respiring tissue and is in the veins. Now because plasma bicarbonate levels rise, pCO_2 declines, and CO_2 diffuses out of the red cells, so that when it enters the gills, the blood has a higher plasma bicarbonate concentration and a lower red cell pCO_2 than blood leaving the tissues. Diffusion of CO_2 from the red cells into the plasma elevates red cell pH, and so the oxygen affinity of the haemoglobin rises and oxygen is taken up at the gill, whilst bicarbonate and CO_2 diffuse outwards.

Table 5.4   Vascular bed of different muscles and diffusion distances. (From Bone, 1978)

Species	Scomber	Katsuwonus	Rat (trained)	Cat
Muscle	red myotomal	deep red myotomal	soleus	soleus
Fibre diameter (μm)	29.6	31	56	54
Mean no. capillaries around one fibre	2.8	4.75	6.39	3.8
Internal diffusion distance (μm)	14.8	15.5	28	27
External diffusion distance (fibre circumference (μm)/ 2 × no. of capillaries around one fibre	16.6	10.25	16.2	22.3

Oxygen and CO_2 transport are complementary, and the whole mechanism is thus efficient enough to satisfy the gas transport requirements of such active fish as tuna. It is probably generally true that all fishes (with a very few specialized swimmers as exceptions) use most oxygen in the red muscle which drives cruising swimming (Chapter 2). In fast-swimming fishes, such as tuna or mackerel, this tissue consumes a great deal of oxygen, and is extremely well-vascularized, having capillary fibre ratios of 1:6 or 1:7 which reduce the diffusion distance of oxygen from the circulatory system to the mitochondria within the muscle fibres. Such ratios and diffusion distances are equal to, or better than those of mammals (Table 5.4). We have already seen that the red colour of the red muscle used for sustained swimming is in part due to a high myoglobin content, used to facilitate oxygen diffusion within the muscle cells. Although sufficient data are not yet available for a variety of fish, it seems likely that the amount of red muscle, and hence (other things being equal), the cruising speed of the fish, is limited by the amount of oxygen that can be acquired at the gills, and here scombroids represent the limiting case.

CHAPTER SIX

OSMOREGULATION AND ION BALANCE

6.1 The osmotic problem

Fishes live in natural waters ranging from nearly pure to hypersaline pools where the water is so dense that they have difficulty in swimming below the surface. Although their skins are as a rule relatively impermeable, they are obliged to have a large gas exchange surface, and there are other permeable surfaces (like the oral and narial mucosae) which are in contact with the water. Also, some water will inevitably be swallowed during feeding.

This means that fish can exploit extra-renal routes of excretion and ion uptake, but since very few fish are isosmotic with the external medium, there will be osmotic gradients across these permeable surfaces. Almost all marine fish have body fluids that are more dilute than sea water, and all freshwater fish have body fluids more concentrated than fresh water (Figure 6.1). In the sea, therefore, most fish will tend to lose water across permeable surfaces (and gain ions), and in fresh water to gain water (and lose ions). It is remarkable that the osmoregulatory mechanisms for coping

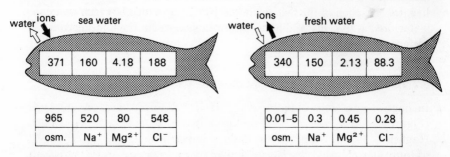

Figure 6.1 Plasma and ambient water molarity and ion content in fresh water and marine teleosts (representative values only). After Pang *et al.* (1977).

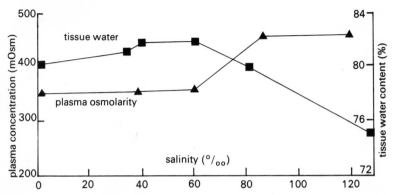

Figure 6.2 Effects of salinity change on killifishes (*Fundulus*). Tissue water content and plasma osmolarity in killifish adapted to different salinities. After Feldmeth and Waggoner (1972).

with these opposite problems can be switched by some fishes as they migrate between the sea and fresh water. Changes in body fluids do indeed occur during such confrontations with different environments and in euryhaline fish, which can adapt to fresh or salt water, like the eel, flounder and salmon, plasma osmolarity in sea water is some 20 % higher than it is in fresh water, mainly as a result of increased plasma sodium. The killifish *Fundulus* holds the record for surmounting the greatest osmotic challenge; Feldmeth and Waggoner (1972) found these fish in hypersaline pools in southern California (128°/$_{oo}$ salinity), but they can also live in fresh water. Between 0–60°/$_{oo}$ NaCl, *Fundulus* can regulate body water and plasma salt concentration; higher salinities are tolerated by accepting tissue water loss around 5 % and increase in blood osmolarity of some 30 % (Figure 6.2).

Before considering problems of osmoregulation in fishes whose body fluids are either more or less concentrated than the water in which they live, let us first look at hagfish, where the body fluids are similar to sea water, and there is little or no osmotic gradient across the permeable surfaces.

6.2 Hagfishes and the origin of the glomerular kidney

In hagfish, there are kidneys of the usual vertebrate mesonephric type (Figure 6.3), although the persistent pronephros remains rather enigmatic. Hagfish blood is isosmotic or slightly hypertonic to sea water; blood sodium and chloride levels are similar to those in sea water, although divalent ions are significantly lower (see Table 6.1). Tissue sodium and

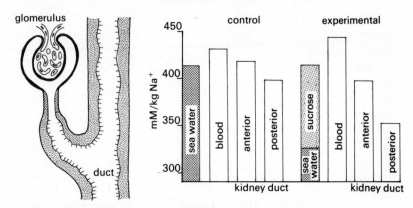

Figure 6.3   The hagfish kidney. Left: single kidney tubule of *Myxine*. Right: sodium levels in control and experimental *Eptatretus* (see text). After McInerney (1974).

chloride levels are high, but even so there is a greater difference between blood and tissue ions than in other fish and osmotic balance between blood and tissues is maintained by high levels of intracellular amino acids. Experiments with tritiated water have shown that hagfish are freely permeable to water, much more so than are other fish where the blood is osmotically different to the medium and it is not surprising that hagfish are unable to tolerate fresh water and are killed by $25^o/_{oo}$ sea water.

There is no doubt that the acraniates arose in the sea, and the situation in hagfish (which are all marine) seems to show that they too have always been marine, and have inherited their 'sea water' blood composition from marine protochordates. Yet lampreys (which either live entirely in fresh water, or have a freshwater larval stage in the life history) are quite different to hagfish in their ionic composition and osmoregulation, as we shall see, and investigations on hagfish urine composition by McInerney (1974) have shown unexpected sodium-concentrating mechanisms in the kidney. Munz and McFarland (1964) concluded that hagfish lacked any mechanism for re-absorbing sodium from the glomerular filtrate (a necessary trick for any animal living in fresh water), and suggested that this was good evidence for the marine origin of the group. But McInerney's careful experiments on hagfish set up in water with low salt content (made isosmotic with sea water by adding sucrose) showed that they responded to low external sodium by increasing renal sodium uptake (Fig. 6.3), and it is difficult to think that this 'surplus' capacity of the sodium recovery mechanism would be needed in sea water. McInerney suggests that it could be considered as a left-over from an original freshwater ancestry. This view would fit with Homer Smith's famous hypothesis of 1932 that the glomerular kidney first arose in fresh water as a device for excreting water. Although the two living agnathan groups are so different from each other that their relationships are hard to ascertain, both could be considered as being originally freshwater, But the alternative view that the glomerular kidney first arose in sea water, as a device to regulate ions by producing a filtrate which could be selectively altered, and that marine fish possessing glomerular kidneys were 'pre-adapted' to enter fresh water now seems more probable, and both hagfish and lampreys are today considered to have arisen in the marine environment. The earliest fossil lamprey, *Mayomyzon* of the Pennsylvanian, is associated with a marine fauna, and *Jamoytius* (Ritchie, 1968) which may be ancestral to both living agnathan groups is also marine. Further investigations of the sodium recovery mechanism of hagfish are obviously desirable.

6.3 Lampreys

All lampreys develop in fresh water, passing several years as filter-feeding ammocoete larvae buried in mud. Some species remain in fresh water after metamorphosis, but others pass down streams and rivers to the sea, where they feed and grow until they return to fresh water to breed. They therefore have similar problems of osmoregulation to such anadromous fish as salmon, and it is very striking (considering the remote affinity of agnathans and teleosts) that both groups adopt similar solutions.

In fresh water, the larval ammocoete and the adult have blood that is more concentrated than the water (Table 6.1); the animals thus tend to gain water. They excrete large amounts of urine, apparently larger amounts than do freshwater teleosts (it is hard to sample urine without stressing the animals, and values for urine production are open to objection on several grounds.) A high rate of urine production enables the animals to get rid of the water which enters via the permeable surfaces, but if there is not to be concomitant loss of ions, there must be selective absorption mechanisms in the kidney so that the ions (or as high a proportion of them as possible) in the glomerular filtrate are reabsorbed as the fluid passes down the nephron to the collecting duct. Lampreys have efficient mechanisms for reabsorbing sodium and chloride but are less efficient at reabsorbing potassium. Isotope studies of ammocoete larvae have shown that the potassium and sodium lost via the urine is replaced via active uptake mechanisms in the gills, and that the animals can maintain a stable blood composition even in very dilute solutions (5–

110

Table 6.1 Composition of plasma and urine in lampreys and in some freshwater fishes. — no data. Values in mM; osmolarity (Osm) in mOsm. From Robertson (1974), Holmes and Donaldson (1969) and Hickman and Trump (1969)

	Lake Huron water	Agnatha — Landlocked sea lamprey Blood ammocoete	parasitic adult	spawning adult	Urine adult in FW	Chondrostei Sturgeon in FW blood	Holostei Amia blood	Teleostei Charr (Salvelinus) blood	urine	Catfish (Ameiurus) blood	urine
Na^+	0.02	103.0	137	136	4.8	155.8	132.5	161	17.4	122	12.2
K^+	0.05	3.4	3.3	5.1	0.99	4.3	2.0	2.8	2.5	2.7	1.61
Ca^{2+}	0.9	2.4	2.2	1.8	—	2.3	5.3	2.05	0.95	—	—
Mg^{2+}	0.25	1.6	2.0	2.7	—	1.47	0.4	0.75	0.55	—	—
Cl^-	0.05	91.0	122.0	112.0	4.7	119.7	119.5	140.6	8.1	110.0	18.0
SO_4^{2-}	2.3	0.1	0.1	0.7	—	0.7	2.2	—	—	—	—
HCO_3^-	1.75	6.0	5.0	5.2	—	—	—	—	—	3.4	0.4
Osm.	—	—	—	241	36	318	—	328.0	36.2	—	—

30 μm Na/l). Here then is the first appearance of extrarenal routes for ion uptake, complementing mechanisms in the kidneys; such mechanisms are more familiar in freshwater teleosts (see below).

In the sea, lampreys face the reverse problem of losing water. However, adult lampreys are only rarely caught at sea (they can usually be seen attached to any basking shark closely examined, but they drop off when the shark is caught), and so there are very few analyses of the blood of marine lampreys. When they are caught as they enter rivers, the marine osmoregulatory mechanism has already changed so that the animals cannot live in sea water; some are able to tolerate 50% sea water (Pickering and Morris, 1970), and it is on such already partially adapted animals that the mechanism of marine osmoregulation has been examined. Pickering and Morris found that, like marine teleosts, marine lampreys drink sea water, thus absorbing water and monovalent ions. These are excreted by specialized 'chloride' cells on the gills, and divalent ions entering the body fluids are excreted by the kidneys. The blood is more dilute than sea water, and although most of the water drunk is absorbed, little urine is produced and excess water is lost extrarenally.

The distant relationship between Agnatha and teleosts suggests that they must have independently evolved similar mechanisms for osmo-regulation, and have had a similar history of secondary return to the sea. Before considering this, let us look first at teleost osmoregulation.

6.4 Teleosts

Teleosts have radiated widely in the sea and in fresh water; many are able to cope either with migrations from one to the other, or to live in estuaries and brackish water of varying salinity. What is more, some are known to have become adapted during their evolutionary history to one environment, and secondarily and relatively recently have entered another (like the fish with aglomerular kidneys that are known in fresh water), so that the group show a wide spectrum of morphological and physiological adaptations to the different osmotic problems these different modes of life incur.

6.4.1 Marine teleosts

Marine teleosts maintain their blood concentration much lower than the surrounding sea water, thus facing osmotic water loss across the permeable surfaces, as do marine lampreys. Homer Smith showed in 1930 that they overcome this by drinking sea water, and absorbing water through the gut wall to make up osmotic water loss; the marine fish he studied drank about 0.5 % of their body weight per hour.

Further investigations showed that marine teleosts differed considerably in their drinking rates. These could be altered by increasing the concentration of the water in which the fish were maintained. In *Serranus*, around 12 % of the body weight was drunk each day in normal sea water, whilst eels drank over twice as much if kept in 2× sea water as in normal sea water. Obviously the relative gill areas differ in different fish (see Chapter 5) and greater water losses are suffered by active fish requiring a large gill area to acquire oxygen.

About 75 % of the water drunk is taken up by the intestine, and since little urine is passed, this process can maintain water balance. But, because water uptake in the intestine is coupled to salt uptake, this method of maintaining water balance incurs the penalty of getting rid of large amounts of sodium and chloride against a concentration gradient; replacing the water problem by an ionic problem (Potts, 1976).

In silver eels in sea water, Kirsch and his colleagues (1981) have been able to show by chronically perfusing the oesophagus that the oesophagus is impermeable to water, but permeable to Na^+ and Cl^- which diffuse across into the blood down their concentration gradients. The fluid entering the intestine is therefore less concentrated than sea water (and nearly isotonic with the blood), and since the intestine *is* permeable to water, water is taken up by the intestine (Figure 6.4). A similar role of the

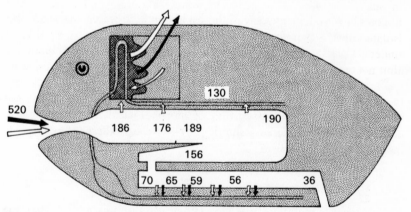

Figure 6.4 Schematic diagram showing osmoregulatory activity of silver eel (*Anguilla*) gut in sea water (eel has been shortened to fit page). Open arrows, water movement; solid arrows, salt movement; numbers in meq Cl^- per l. After Kirsch *et al.* (1981).

oesophagus in osmoregulation is likely in other marine fish; for instance, experiments on bass have shown that acid secretion by the stomach is inhibited if perfused with sea water, whereas 50% or 30% sea water permits normal secretion.

Early experiments with perfused fish heads showed that the salt taken up in the alimentary canal was actively transported from the blood to sea water in the head region (actually across the gills), but it was not until radioactive tracers were used to examine salt fluxes across fish (mainly in a long series of experiments by the distinguished French physiologist Jean Maetz) that a completely unsuspected massive salt *influx* was discovered to take place across the gills. Not surprisingly, this discovery completely altered ideas of mechanisms of salt exchange across the gills, and it became clear that total salt influx was 5–10 times greater than that resulting from drinking sea water and that much more salt was lost across the gills than had hitherto been supposed.

Figure 6.5 illustrates the salt balance of a marine teleost disclosed by tracer experiments. The rate of movement of salt across the marine teleost gill is very much greater than that across any other epithelium (Potts, 1976) and across the chloride cell apices concerned with ion movements, it is continuously about the same as that across the squid giant fibre membrane at the height of the action potential. How are these massive fluxes brought about? This problem has still not been clarified entirely.

Two kinds of experiments have contributed to its solution. By loading fish with labelled Na^+, efflux measurements can be made when the fish are in sea water, or in variously modified solutions, and (with greater difficulty) it is possible to measure influx rates in a similar way. Secondly, isolated perfused gills can be used to measure transepithelial potentials under different conditions. The first approach showed that in some fish (but not in all) sodium efflux depended upon sodium concentration in the

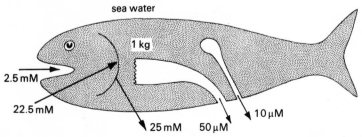

Figure 6.5 Salt balance in a marine teleost revealed by tracer experiments. After Potts (1976).

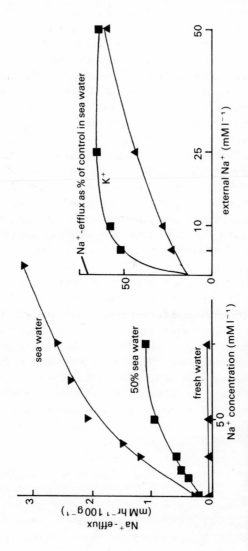

Figure 6.6 Sodium efflux from the flounder *Platichthys* under various experimental conditions. Left: relation between Na$^+$-efflux and Na$^+$ concentration of external medium in fishes adapted to sea water and diluted sea water. Right: relation between Na$^+$-efflux and Na$^+$ and K$^+$ concentration in the external medium. Note that low external K$^+$ is more effective than low external Na$^+$ in stimulating Na$^+$ efflux. After Motais *et al.* (1966) and Maetz (1969).

medium (Figure 6.6), and that external potassium concentration also affected sodium efflux. Since the gills contain high concentrations of Na–K activated ATPase, Maetz (1969) proposed an active K–Na exchange process (driven by the ATPase) at the gill, the potassium entering simply diffusing outwards along the concentration gradient into the sea water again. In addition to this *active* process, a *passive* one-to-one exchange of external for internal Na ions was suggested, resulting from the high external Na concentration, this exchange diffusion process being mediated by the same transport system as the Na–K exchange, but requiring no energy.

Experiments on isolated perfused gills showed that the gill in most marine teleosts is about 20 mV positive to the external solution, and this is probably the consequence of the much greater permeability of the gill to sodium than to chloride, which facilitates the movement of sodium outwards along the electrochemical gradient. Some contribution to the positive potential of the gill in sea water is also made by the Na/K pump, operating in parallel to the diffusion potential. As yet, the large chloride fluxes seen in marine teleost gills have not been explained (see the review by Maetz, 1974).

6.4.2 Salt balance in fresh water

In fresh water, water enters across all permeable surfaces, and there is a large concentration gradient favouring diffusion of salts across these surfaces, so the freshwater fish faces quite different problems to the marine teleost. Drinking rates are low, and the water influx is met by excreting dilute urine in large amounts. The urine is much less concentrated than the plasma (see Table 6.2) for not only is little water resorbed from the glomerular filtrate as it passes along the tubule, but salt is very efficiently absorbed. In the pike and suckers studied by Hickman (1965), over 99.9% of the filtered Na^+ and Cl^- were resorbed, and from filtrate osmolarities of 220–320 mOsm, the urine is diluted to 20–80 mOsm by ion resorption. This process maintains water balance, but salt is lost across the permeable surfaces (mainly across the gills) and freshwater teleosts have developed a very efficient high-affinity salt uptake system at the gills to maintain salt balance. For example, Krogh (1939) showed that goldfish which had been kept in distilled water for some weeks were able to accumulate sodium chloride activity from solutions as dilute as 10^{-4}–10^{-5} M, and eels can accumulate sodium from solutions containing only 0.05 M NaCl. Various kinds of evidence show that sodium and chloride uptakes are independent, so that selective blocking of the uptake of one does not affect the uptake of the other. As Maetz (1974) points out, 'In order to preserve electrostatic neutrality in such single ion transfers, the body surfaces must contain a pair of ionic exchange systems capable of operating independently of each other'.

After the early divided-box experiments of Smith (1929), it became clear that the main route for nitrogen excretion in teleosts was via the gills, and that this was mainly in the form of ammonia. Branchial excretion of ammonia not only provides a convenient and simple way of getting rid of the waste nitrogen resulting from transdeamination pathways, but was long supposed to be involved in sodium uptake in fresh water. Krogh (1939) suggested that Na^+ was exchanged for NH_4^+, and that Cl^- was exchanged for HCO_3^-, and when fish are in a steady state with the water in which they are kept, the rates of ammonium loss and Na^+ uptake are indeed similar. But marine teleosts also excrete ammonia across the gills, whilst at the same time they excrete rather than absorb Na^+, and injection of acetazolamide (which blocks Na uptake) has no effect on NH_4^+ excretion. Clearly then there is not an obligatory exchange of NH_4^+ for Na^+, and when Na^+ is not taken up, NH_4^+ must be excreted accompanied by an anion, probably HCO_3^-. NH_4^+ movements across the gill cannot

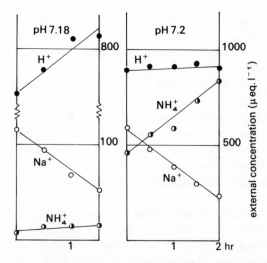

Figure 6.7   Sodium uptake in goldfish. Measurements for two goldfishes of external Na^+, titrable acidity, and total ammonium. Note that fish on the left shows $Na^+ - H^+$ exchange; fish on right shows $Na^+ - NH_4^+$ exchange. After Maetz (1974).

therefore alone account for Na^+ uptake, as Krogh suggested, and it now is clear that Na^+ is also exchanged for H^+. Figure 6.7 shows the situation in two goldfish, one of which excreted few NH_4^+ ions and where Na^+-H^+ exchange was the predominant mechanism of Na^+ uptake; in the other H^+ secretion was minimal and Na^+ uptake was by the $NH_4^+-Na^+$ exchange mechanism. The sum of the NH_4^+ and H^+ ion movements is well correlated with the net Na^+ uptake.

The hypothesis that Cl^- is exchanged against HCO_3^- has withstood experimental attack, and follows from the richness of the enzyme carbonic anhydrase in gill tissue and from the fact that the gill is the major route of CO_2 excretion. For obvious reasons, experiments on Cl^- uptake are usually made in Na^+-free solutions and vice versa, but when both ions are present in the external medium, Cl^- entry is facilitated.

6.4.3 Chloride cells

It is remarkable that the same cell type provides the special cells concerned with salt uptake processes in fresh water, and those excreting salt in the sea. Special acidophil cells on the eel gill were described nearly 50 years ago, and because they were more abundant on the gills of eels adapted to sea water than in those adapted to fresh water, they were thought to be responsible for salt exchange at the gills. Later ultrastructural and histochemical work has confirmed the view that in sea water such chloride cells (so called because they contain chloride, but 'ionocyte' is really a more descriptive term) are concerned with salt extrusion, and in fresh water, with salt uptake. How do they work?

The structure of chloride cells is as in Figure 6.8. The basal and basolateral parts of the cell are extensively infolded, forming a system of tubes extending almost to the apex of the cell, which projects into the surrounding water. This smooth tubular system represents extracellular space, and is in close contact with the blood of the gill; around it in the chloride cell are large numbers of closely packed mitochondria. The walls of the tubular system contain high concentrations of Na^+-K^+ activated ATPase (concerned with Na transport) and at the ultrastructural level these are seen as regular particle arrays on the tubular-system membrane. Inward movement of NaCl through the chloride cells of freshwater fish takes place through the cell apices, exposed to the water. Carbonic anhydrase is present in large amounts, catalysing the hydration of CO_2 in the cell ($CO_2 + H_2O = H_2CO_3 = H^+ + HCO_3^-$).

Na enters as H^+ ions are excreted, and as Cl^- enters so HCO_3^- is

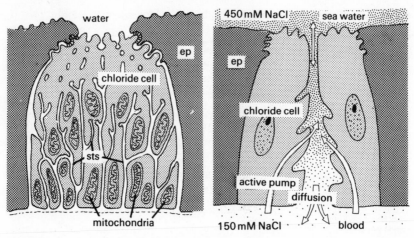

Figure 6.8   Chloride cells. Left: schematic ultrastructure showing apical villi, junctions with adjacent epithelial cells (ep), and smooth tubular system (sts). Right: simple scheme for secretion into sea water between 2 adjacent chloride cells. After Sargent *et al.* (1978); Sardet *et al.* (1980).

excreted. In sea water, the chloride cell is required to secrete salt against a high concentration gradient, and a recent suggestion by Sargent *et al.* (1978) provides a simple scheme whereby this process may take place. Sargent and his colleagues pointed out that in marine fish, or in fish adapted to sea water, there are many chloride cells, and thus many will lie adjacent to other chloride cells. They suggest that the high density of sodium pumps facing the spaces between adjacent chloride cells produces a concentration of NaCl in these spaces higher than that in sea water, so salt actually diffuses outwards through the junctions between adjacent chloride cells down a concentration gradient into sea water. Of course, much will also diffuse back into the blood and be recycled by the chloride cells. This scheme, summarized in Figure 6.8, certainly has the merit of simplicity, and awaits a direct test by examination of frozen sections by electron microprobe X-ray analysis. Definitive proof that the chloride cells are the sites where salt is secreted has recently been given by Foskett and Scheffey (1982), who have shown that negative current density peaks are correlated with chloride cell apices visualized with a fluorescent ionophore.

Similar chloride cells are found in marine elasmobranchs, but have not been reported from freshwater members of the group; they have also been found in Dipnoi.

6.4.4 The kidney and salt balance

So far we have mainly considered extra-renal routes of excretion and ion balance in fish; the role of the kidney itself has hardly been mentioned. Fish kidneys are typical mesonephroi (Figure 6.3) retaining a segmental structure, most evident in their blood supply, which is essentially venous (Figure 6.9), with a renal portal system and operating at low systemic blood pressures not above 20 mm Hg. A particularly interesting feature of the fish kidney is that the nephrons in different fish are extraordinarily diverse, so that the function of the different nephron segments can be inferred by comparing different fish. Perhaps the most striking modification of the nephron occurs in some marine teleosts (23 species of 7 families at the present count) where the glomerulus is reduced or even completely lost. Study of such aglomerular kidneys has been fruitful for renal physiology. In this way tubular secretion was first demonstrated, and insulin was found to be the tracer of choice for the glomerular filtration rate (GFR).

As we should expect, urine production and indeed kidney function are different in freshwater and marine teleosts. Freshwater teleosts produce between 0.1 and 1.4 ml of urine hr^{-1} 100^{-1} g body wt., whilst in most marine species, urine production is around 1/10 of these values. The

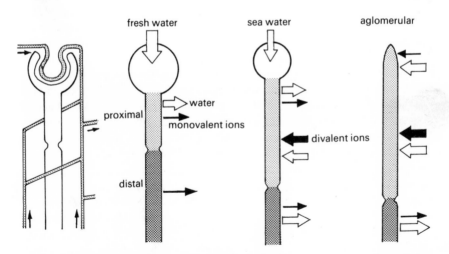

Figure 6.9 The teleost nephron. Left: blood supply (venous, except to glomerulus); right: three functional types in freshwater and marine teleosts. Note development of proximal region in marine teleosts. In aglomerular marine teleosts, the distal region includes also the collecting ducts and bladder. After Lahlou (1981).

relative clearance of water via the kidney is up to 30% in fresh water, and practically nil in sea water, as the urine in fresh water is much more dilute than the plasma; in sea water it is more or less isosmotic with the plasma. Wide simultaneous variations in GFR and urine production take place in normal fish without change in urine concentration, and it seems that recruitment of glomeruli under varying conditions plays an important role in the function of the kidney. Even small variations in systemic blood pressure (since it is always low), can 'shut down' glomeruli, and coupled with linked changes in tubular absorption greatly change urine production without change in urine concentration. Measurements of GFR in single nephrons of the trout, for example, have shown that in fresh water, 45% of the nephrons are filtering, whereas in sea water, only 5% filter. Obviously, such a flexible system has the great merit that it allows more or less automatic regulation of urine production as euryhaline fish move from the sea into fresh water or vice versa, and the osmotic loads change.

6.4.5 Tubular structure and function

Although there is a wide spectrum of tubular structure in different fish, well described by Hickman and Trump (1969), three basic types are all we need recognize here (Figure 6.9). The proximal segment next to the glomerulus (where it occurs) is in two parts, one equivalent to the mammalian proximal segment, the other concerned with the secretion of divalent ions; the distal segment being concerned with water and monovalent ion absorption, and in elasmobranchs, with urea absorption. In fresh water, monovalent ions are conserved by tubular absorption, whilst in marine fish, divalent ions are secreted by the tubule. Although Hickman and Trump remark that 'it would scarcely do the marine teleost kidney justice to consider it purely a magnesium sulfate pump', this is certainly one of its main roles. Mg^{2+} and SO_4^- enter the plasma via the sea water drunk (sea water contains around 50 mM Mg^{2+} and 25 mM SO_4^{2-}) and are excreted exclusively through the kidney. Na^+, Cl^- and other ions are also excreted, but extra-renal routes are more important for these ions. As soon as marine euryhaline fish, such as the southern flounder *Platichthys flesus*, enter diluted sea water, Mg^{2+} secretion by the distal tubules drops dramatically, without notable change in plasma Mg^{2+} concentration (Figure 6.10), and then, more slowly, the tubules become less permeable to water, and Na^+ is more completely absorbed so that the urine osmolarity declines.

In elasmobranchs, though not in *Latimeria*, 95% of the urea in the

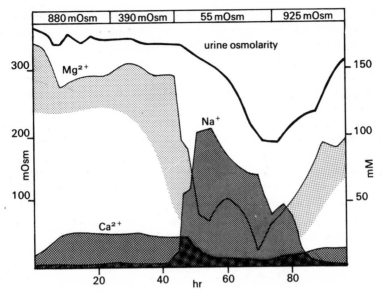

Figure 6.10 The effects of changes in external salinity on kidney function in the euryhaline southern flounder (*Platichthys*)—salinity changes (mOsm) at top. Note rapid change in urine Mg^{2+} concentration and increase in Na^+ as salinity drops. After Hickman and Trump (1969).

glomerular filtrate is resorbed by the distal tubule perhaps by a countercurrent system where the initial thinner part of the distal tubule is applied to the proximal tubule (Figure 6.11).

After the distal region of the tubule, collection ducts of differing sizes in different species pass the urine to a urinary bladder, which in some species at least, plays an important role in modifying the ionic content of the urine.

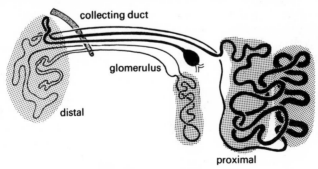

Figure 6.11 Single nephron of ray. After Boylan (1972).

In *Lophius* and *Platichthys* for example, water is resorbed when the fish are kept in sea water, and in trout in fresh water ions are resorbed and conserved. These processes are under the control of the pituitary hormone prolactin.

6.5　Osmoregulation in elasmobranchiomorphs

In elasmobranchs and Holocephali, which are today mainly marine (though there are elasmobranchs which migrate into fresh water, and a few which are exclusively freshwater), blood ionic composition is similar to that of marine teleosts, but blood osmolarity is close to that of sea water (Table 6.2). It is so because in addition to the usual ions, the blood and tissue fluids contain large amounts of low molecular weight nitrogenous solutes. Most abundant of these is urea (known to be at high concentration in elasmobranchs so long ago as 1858); urea is usually present at around 0.4 M. Various methylamine substances such as trimethylamine oxide (TMO), betaine, and sarcosine; and some free amino acids like taurine and β-alanine are also in fairly high concentration, in total around 0.2 M. Over half of the blood osmolarity is thus due to these nitrogenous solutes, and as a result, blood osmolarity is close to that of the sea water, and the osmotic load across the gills is small. It seems likely that marine elasmobranchs are always slightly hyperosmotic to the sea water in which they live, for example dogfish in Plymouth circulation sea water (osmolarity $1154 \, \text{mOsm} \, 1^{-1}$) had serum at $1243 \, \text{mOsm} \, 1^{-1}$. The gills are permeable to water, so that under normal conditions there will be a constant slight influx of water, excreted as urine that is more dilute than the serum or sea water (Table 6.2). Urea and TMO are efficiently resorbed from the glomerular filtrate, and most of the 1–2 % of total daily urea loss occurs across the gills. Thiourea (differing from urea only by the substitution of a sulphur for an oxygen atom) is much less efficiently resorbed by the urea transport mechanism of the kidney, but it is lost at similar rates as urea across the gills. Thus either there is a different urea transport mechanism at the gill to that of the kidney, or the low rate of urea loss across the gills may be simply the result of obstruction of urea outflow by the inward flow of water into the hypertonic serum. By placing sharks in media more concentrated than normal sea water the rate of urea loss across the gills is much increased.

6.5.1　Urea and proteins

The retention of urea and hence the reduction of the osmotic load across the permeable surfaces of marine elasmobranchiomorphs means not only

Table 6.2 Composition of plasma and urine in some marine fishes. Values for seawater representative. Values in mM. From Pang et al. (1977) and Griffith and Pang (1979)

| | Sea water | Agnatha | | Elasmobranchiomorpha | | | | Coelacanth | | Teleostei | | | |
| | | Myxine | Eptatretus | Chimaera | Hydrolagus | Squalus | Squalus | Latimeria | Latimeria | Fundulus | Muraena | Paralichthys | Lophius |
		blood	urine	blood	urine	blood	urine	blood	urine	blood	blood	urine	urine
Na$^+$	470	487	553	338	162	296	240	197	184	183	212	59	11
K$^+$	10	8.4	11	11.7	7.8	7.2	2.0	5.8	9	4.8	2.0	3.4	2.0
Ca^{2+}	10	4.8	4	4.3	17	3.0	3.0	4.8	2.0	2.3	3.9	11	7
Mg^{2+}	54	9.3	15	6.1	69	3.5	40	5.3	30	2.1	2.4	78	137
Cl$^-$	548	500	548	353	268	276	240	187	15	146	188	124	132
HCO$_3^-$	—	7.2	9	2.6	—	—	—	9.6	—	13.3	—	—	—
PO$_4^{3-}$	—	0.4	7	—	25	2.4	33	5.1	38	5.3	—	11	2.0
SO$_4^{2-}$	28	3.7	9	5.2	26	3.1	70	4.8	104	—	5.7	28	42
urea	—	2.8	—	332	52	308	100	377	384	4	9.1	—	0.6
TMAO	—	—	—	0.0	—	72	10	122	94	—	—	—	13
amino a.	—	—	—	—	—	11.6	—	16	—	8.5	—	—	—
Osmol. (mOsm)	1011	969	—	1046	820	998	800	932	962	363	—	295	406

that there must be urea resorption mechanisms in the kidney, but also poses another quite different problem. At concentrations around 0.5 M, urea disrupts proteins in most animals. Collagen, haemoglobin, and many enzymes in mammals are denatured by such concentrations of urea. Either elasmobranchiomorphs have in some way modified their proteins to resist the effects of urea, in the same way that puffer fish (*Tetrodon*) have modified their sodium channels to resist the effects of the tetrodotoxin (TTX) which rapidly poisons those of other animals including man, or they have protected the proteins by some other trick.

Some elasmobranch proteins, like their haemoglobin, *are* resistant to urea, but recent work by Yancey and Somero (1978, 1979) on several elasmobranch enzymes has shown that the situation is here more complicated, and that the enzymes are actually protected against urea inhibition by the presence of the other nitrogenous solutes such as TMO (see also Altringham *et al.*, 1982). Maximum counteracting effects of the methylamine compounds and urea on several skeletal muscle enzymes (such as creatine kinase, lactate dehydrogenase, and pyruvate kinase) were found at urea: methylamine ratios of 2:1, the ratio that is actually found in elasmobranchs, Holocephali, and in *Latimeria*, of which more later. This striking discovery showed how by using waste products these fish have been able to cope with the problems of osmoregulation in sea water in an economical way; the same kind of strategy as the more wasteful one used by several invertebrates, which increase their serum osmolarity with high concentrations of free amino acids.

The ionic composition of marine elasmobranch sera is not unlike that of marine teleosts, and very different to sea water; for example Mg^{2+} and SO_4^{2-} are at much lower values than sea water. These divalent ions are excreted by the kidney, and values in urine are much higher than they are in the glomerular filtrate.

6.5.2 *Extrarenal salt excretion and the rectal gland*

Like the teleost kidney, that of the elasmobranch is an effective magnesium sulphate pump. But although Na^+ and Cl^- are at lower concentrations in the serum than they are in sea water, blood and urine values for these ions are similar, so that extra-renal routes for NaCl excretion must exist. One such route was unexpectedly found by Burger and Hess (1960), who showed that the rectal gland (Figure 6.12) secreted concentrated NaCl solution, and it has since been found that the cells of the gland have the same high Na-K-activated ATPase enzyme activity and ultrastructure as

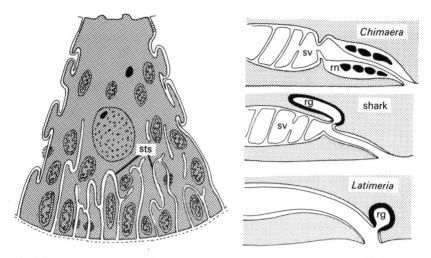

Figure 6.12   The salt-secreting rectal gland. Left: schematic ultrastructure of *Latimeria* rectal gland cell (cf. Figure 6.8). sts, smooth tubular system. Right: rectal glands (rg) in elasmobranchiomorphs and *Latimeria*. In Holocephali, the rectal gland tissue forms nodules in the rectal wall (rn). After Lemire and Lagios (1979); Lagios (1979).

chloride cells of fishes and salt glands in other animals. The rectal gland is, however, not the only route of NaCl excretion in elasmobranchs.

Tracer experiments by Payan and Maetz (1973) have shown that 2/3 of the total sodium loss in elasmobranchs occurs across the gills. In similar experiments on the pyjama shark (*Poroderma*) Haywood (1975) was able to show that even after ligation of the rectal gland and the urinary system the shark could regulate sodium and chloride. He found that, as in teleosts, chloride cells were present on the gills, and these are presumably responsible for NaCl excretion.

6.5.3 Freshwater elasmobranchs

A number of elasmobranchs, including sawfishes (*Pristis*) and the bull-shark *Carcharinus leucas*, are euryhaline, able to live for long periods in brackish estuaries and even in fresh water, as well as in the sea; naturally these have long interested zoologists—such fish were first studied by Homer Smith in the Perak river of Malaysia. More recently, Thorson and his colleagues (1967) have studied *C. leucas*, which swims up the San Juan river into Lake Nicaragua (some 350 km from the sea). In all the euryhaline species caught in fresh water, blood levels of urea, sodium and chloride are reduced as compared to the values for the species in sea water (for example urea is reduced to 25–35 % of the marine value), but of course, the fish are still much above the osmolarity of the water, and produce a copious flow of dilute urine to deal with the water influx across the gills.

Apart from these euryhaline species, there are also a few stingrays in the family Potamotrygonidae which live in the Orinoco and Amazon drainages up to 4500 km from the sea, and never enter sea water. Fossil stingrays of the same family have been found in Tertiary deposits of the Rio Parana basin, so it seems probable that the family has lived in fresh water for millions of years.

These stingrays have done just what one would advise an elasmobranch to do in fresh water: abandoned urea retention. They have also reduced the ionic content of the blood as compared with marine rays (Table 6.2) and so the blood is strikingly similar to that of freshwater teleosts. When Thorson and his colleagues (1967) first discovered the virtual absence of urea in *Potamotrygon* blood, it was not clear whether return to sea water could evoke increased urea production via the ornithine cycle, i.e. whether return to the normal elasmobranch situation was possible. Later work by Griffith *et al.* (1973) has shown that these rays are incapable of osmoregulation even in dilute sea water, and cannot retain urea. They have rectal glands, but these are reduced and non-functional. So *Potamotrygon*, like the freshwater teleosts, can take up Na^+ at the gills, presumably via the chloride cell system.

Obviously, these 'natural experiments' in elasmobranch osmoregulation have interesting implications for theories about the origin of the urea retention system of marine elasmobranchs, but before considering these, let us look at *Latimeria*.

6.5.4 Latimeria—independent urea retention?

Latimeria blood (Table 6.2) is similar in composition to that of marine elasmobranchs, containing similar amounts of urea, TMO, and amino acids as well as inorganic ions. Total osmolarity is less than sea water, rather than just above it as in elasmobranchs, but there are grounds for supposing that *Latimeria* may live in caves and fissures in the basaltic rock of the Comoro Islands, and that aquifers opening into these lairs produce low-salinity environments. So although hypo-osmotic to normal sea water, *Latimeria* may be hyperosmotic to the water it lives in, and thus faces a slow inward water flux across the relatively small gills, just like marine elasmobranchs. Urine analyses by Griffith and Pang (1979) (Table 6.2) show that like marine elasmobranchs, *Latimeria* excretes divalent ions, at levels higher than they are in the blood. Na^+ in the urine is, as in sharks, at the same level as in the blood, though Cl^- is lower, but unlike the elasmobranch, there seems to be no mechanism for tubular urea resorption, since urea levels in the urine are the same as those in the blood. *Latimeria* must lose urea at significant rates via the kidneys, and perhaps across the gills also, though these have a small area. Chloride cells are apparently few in the gill epithelium, but there is a rectal gland of similar ultrastructure (perhaps of different origin) to that of marine elasmobranchs (Figure 6.12).

Intensive investigation of *Latimeria* in the last few years has demonstrated several similarities to elasmobranchs in the arrangement of the tissues, in addition to the common urea retention. For example, the adenohypophysis is similar in the two groups, as is the structure of the pancreatic islet tissue. Points such as these have led some workers (e.g. Lagios, 1979) to suppose that the coelacanths and the elasmobranchs are sister groups, but there is no doubt that this view is mistaken, and that the coelacanths occupy the systematic position outlined in Chapter 1. The reader is advised to turn to Lagios' account, and its rebuttal by Compagno (see Lagios, 1979) to see how easy it is to be led astray by concentrating on just a few features of anatomy, histology and biochemistry.

Griffith and Pang (1979) point out various possible ways of looking at the unusual common feature of urea retention in *Latimeria* and elasmobranchs. They conclude that urea retention has evolved independently in the two groups, a conclusion much strengthened by the strange crab-eating frog *Rana cancrivora*. This frog lives in estuaries, and maintains blood osmolarity close to the environment with high levels of urea. In this case, urea is not resorbed by the kidney tubules, but *is* resorbed across the wall of the urinary bladder. Since it is difficult to imagine that this amphibian should also be placed in close relationship with the elasmobranchs, it must be concluded that it has invented urea retention independently, and this seems much the simplest view of the elasmobranchs and *Latimeria* also.

6.5.5 The 'efficiency' of urea retention

Before leaving these two groups we may ask, with Griffith and Pang, what are the advantages of choosing urea retention? It seems clear that in both, urea retention arose under the same circumstances as it has in *R. cancrivora*, that is, as an osmoregulatory mechanism to cope with the marine environment. Why did petromyzonts and actinopterygian fishes opt for a different mechanism—i.e. being hypo—osmotic to sea water? Although this is in one sense an improper kind of question (one might as well ask why elasmobranchs have not invented swimbladders to make themselves neutrally buoyant instead of using liver oils) it does raise other more fruitful questions. For instance, we might go on to ask which method of osmoregulation is likely to be least costly in terms of ion-pumping, and although the calculations are 'order of magnitude' ones only, Griffith and Pang suggest that urea retention is likely to be some 20 times more economical in terms of mM ATP $kg^{-1}h^{-1}$ needed for sodium excretion by a marine teleost on the one hand, and for urea synthesis and limited sodium excretion by a marine shark on the other. Yet the solution of urea retention to the problem of marine osmoregulation requires several features which are not typical of all fishes. First, of course, there must be a complete ornithine-urea enzyme cycle. Probably this was present in ancestral gnathostomes, though absent in lampreys, and probably secondarily incomplete or inactive in most teleosts.

Secondly, the urea produced must be retained, and this will be easier for fish that are large, relatively sluggish, and which have internal fertilization. Most urea lost will be across the gills and other permeable epithelia, relatively smaller in large inactive fish than in small active fishes. The surface/volume ratio will be greatest in the embryonic stages, and it will be difficult for such stages to produce enough urea to counteract its loss across the permeable surfaces. So sharks and coelacanths were better able to adopt urea retention than smaller teleosts with external fertilization. It is

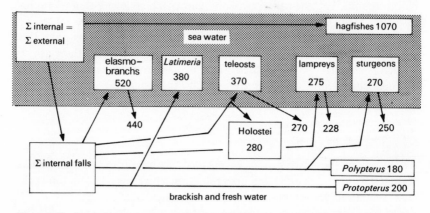

Figure 6.13 Evolutionary sequence of major fish groups. Numbers represent sum of Na^+ and Cl^- in plasma (i.e. approximately total ion content). After Lutz (1975).

perhaps significant that the most active sharks, like the lamnids, which have to have large gill areas to acquire the oxygen they need to sustain their large amounts of red muscle (Chapter 3), are not only large, but their embryos are large at birth.

6.5.6 Plasma ion content and evolutionary history

Different fish groups have a characteristic ionic composition, ranging from myxinoids isosmotic with sea water to the dipnoi and polypteroids whose plasma is the most dilute, and this has naturally made zoologists ponder whether plasma ionic composition can be explained in terms of the evolutionary history of the groups. Lutz (1975) collected values for the Na^+ and Cl^- (the major ionic osmotic constituents) content of the plasma in different groups (Figure 6.13). A good case could be made that the lowest values represented the longest history of freshwater evolution, and that where a secondarily marine phase had developed, the difference between the marine and freshwater plasma concentrations was least in those groups which had most recently re-entered the sea. Lampreys do not fit too well into this scheme, but as only limited evidence is available for plasma composition in sea water, and it seems likely that their migrations into the sea are of relatively recent origin, the freshwater values are probably a safer guide to the evolutionary history of the group.

CHAPTER SEVEN

FOOD AND FEEDING

Adult fish feed in a variety of ways, ranging from sieving phytoplankton or grazing algae to catching other fishes, and as we might expect, some with peculiar diets have devised very specialized methods of feeding. Others

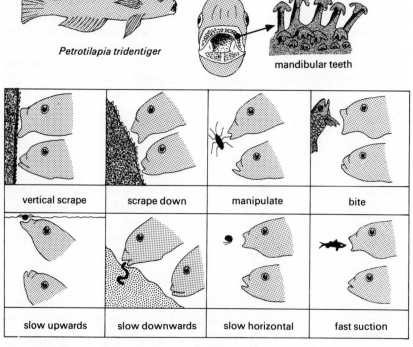

Figure 7.1 Versatile feeding methods of the cichlid *Petrotilapia*. The 3 slow techniques involve slow controlled suction. After Liem (1980).

have a more general diet—for example the blue shark (*Prionace*) feeds on dead whales, fish, cephalopods, adult ascidians, gastropods and crabs! Interestingly enough, as Liem (1980) has shown, some specialized feeders can successfully turn their adaptations to other uses, although apparently highly specialized for one particular diet and mode of feeding. The cichlid *Petrotilapia* of Lake Malawi is seemingly specialized for scraping algae from rocks with its trifid teeth (Figure 7.1), but Liem found that it was in fact capable of feeding in no less than 8 different ways, each involving a particular pattern of jaw muscle activity. For instance, *Petrotilapia* bites fins and scales from other fishes, collects food floating on the water surface, sucks loose prey from the bottom, and catches small fishes in midwater. Such remarkable versatility in the operation of what seems at first sight to be a morphology adapted for a single purpose may be uncommon, but Liem's study serves as a warning when attempting to work out function from morphology alone. It would be hard to guess, if the habit had not been observed, that other cichlids feed by shamming death on the bottom and so decoying other fishes, or that others, such as *Haplochromis compressiceps*, as well as devouring small fishes, suck the eyes out of other fish! These and other modes of cichlid feeding are discussed in a fascinating study of the cichlids of the African Great Lakes by Fryer and Iles (1972). Before looking at the feeding mechanisms and diet of adult fishes, we first glance briefly at the food and feeding of fish larvae.

7.1 Food of larval and young fishes

In the sea, copepods began to diversify and become dominant forms in the Lower Mesozoic, and today they are the largest group of protein producers in the entire world. In fresh waters, cladocerans and rotifers are dominant in the zooplankton. This immense source of conveniently small morsels of protein is ideal for fish larvae—most kinds of bony fish larvae feed on planktonic crustaceans, whatever they eat as adults. In the seas around the British Isles, larvae of all the well-studied species (clupeids, gadoids, mackerel and flatfish) feed largely on copepods.

Most fish larvae show a regular rhythmic swimming pattern, where a few tail beats alternate with periods of rest; this pattern allows economical search of the largest possible volume of water for food, and is also typical of other predators (like *Sagitta* and many medusae). Hunter (1972) examined the feeding behaviour of anchovy larvae (Figure 7.2) and found that the volume of water searched depended simply on the size of the larva, according to the relation (litres searched per h) $\propto$ (larvae length in cm)^3.

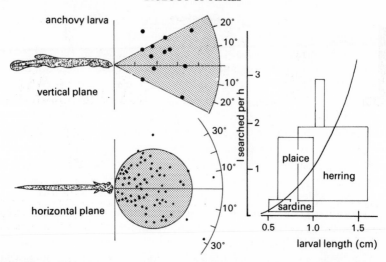

Figure 7.2   Feeding of anchovy larva (*Engraulis mordax*). Left: positions of prey (dots) seen and caught by anchovy larvae in vertical and horizontal planes. Right: relationship between larval length and litres water searched per hour in different species. After Hunter (1972).

The volume searched obviously depends on the size of the visual field (Figure 7.2) within which food particles are seen, and some fish larvae have enlarged this visual field, thus making themselves more effective hunters. Weihs and Moser (1981) have shown that the remarkable stalked eyes of the larva of the stomiatoid *Idiacanthus* (Figure 7.3) allow it to observe a volume over 80 times that of a similar larva with fixed non-rotatable eyes. *Idiacanthus* is a special case, but significant improvement in the field can be obtained by protruding the eyes on a much smaller peduncle, as do many myctophid larvae (Figure 7.3).

In the North Sea, plaice larvae feed almost exclusively on the larvacean *Oikopleura*, and although these fragile tunicates are soon digested in the stomach of the larva, Shelbourne (1962) found that their faecal pellets are

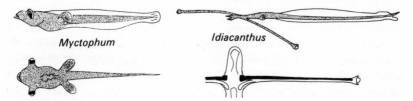

Figure 7.3   Mesopelagic fish larvae with extended eyes which increase area searched for food. After Weihs and Moser (1981).

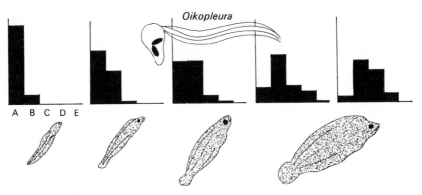

Figure 7.4   Food of plaice larvae in the North Sea. As the larvae grow, they capture larger *Oikopleura*. Histograms show overall length of *Oikopleura* (cm) eaten by larvae at different stages. A: 0.05–0.28; B: 0.33–0.55; C: 0.61–0.83; D: 0.88–1.10; E: 1.15–1.38. After Shelbourne (1956, 1962).

not. Since the size of the faecal pellets is related to the size of the larvacean, measurement of the pellets allowed Shelbourne to calculate the size of the *Oikopleura* that the larvae were feeding on. In this ingenious way, he showed that as the larva grew it could eat larger and larger *Oikopleura* (Figure 7.4) in the same way that copepod feeders take larger and larger copepods as they grow. Evidently the early life histories of bony fishes evolved as did the size and habit of their small prey organisms. The kind of changes in feeding during larval life are well shown by the Caspian roach (*Rutilus r. caspicus*). After the yolk sac has been absorbed, the post-larval to metamorphic stages (5.0–8.5 mm long) feed on *Volvox*, rotifers, and copepod nauplii. The fry up to 12.0 mm have protrusible jaws and capture cladocerans and other planktonic crustaceans, whilst the young roach (15 mm or so) pick insect larvae from the bottom. As adults, the roach feed chiefly on molluscs (Nikolsky, 1963).

Apart from the special case of the large ammocoete lamprey larva, we know of no fish larvae that filter feed on phytoplankton. Instead, all snap up particles one by one, like the anchovy, hunting rather than sieving. Perhaps this is simply a reflection of the fact that successful filter feeding requires a relatively high density of particles to graze upon, and that for small larvae it is better to catch small prey organisms individually. Although filter-feeding anuran tadpole larvae like those of *Xenopus* are successful, there seem to be no fish equivalents. The peculiar leptocephalid larvae of eels and related groups bear teeth inclined forwards from their insertions (Figure 8.7) and it is just possible that this curious arrangement (paralleled on a much larger scale by some ramphorhynchid pterosaurs) may be related to some kind of filter feeding method, though it has also been suggested that leptocephali simply absorb dissolved nutrients.

7.2 Adult fishes

7.2.1 Plankton feeders

The protochordate ancestors of fishes were certainly, as adults, suspension feeders, collecting phytoplankton and detrital particles from the water much as does the lamprey ammocoete larva, but moving the water they filtered by ciliary rather than muscular action. Mallatt (1981) compares the organization of the filtering mechanism in amphioxus and ammocoetes and concludes that they are of common descent. Very few adult fishes live by filtering phytoplankton. The Peruvian anchovy (*Engraulis ringens*) and the Atlantic menhaden (*Brevoortia*) subsist largely on phytoplankton, collecting particles down to 13–16 µm with their elaborate gill rakers. Certain fishes also have epibranchial organs that seem to be involved in microphagy. These are paired pharyngeal diverticula supported by branchial elements that open just behind the gap between the 4th and 5th gill arches. In the thread-fin shad (*Dorosoma pretense*) for instance, the inner series of gill rakers on the lower branch of the fourth gill arch and those of the fifth form a funnel that leads to the opening of the epibranchial organ, which has muscular walls and is lined with mucus-producing cells. It seems that the function of the epibranchial organs is to gather and compact food particles in boluses of mucus that are squeezed into the oesophagus. Moreover, fishes with highly developed epibranchial organs (which besides *Dorosoma* include an osteoglossid (*Heterotis*), the milkfish *Chanos*, and an anchovy (*Cetengraulis*)), feed on the finest phytoplankton and mud particles (Nelson, 1967). These fishes also have a triturating 'gizzard', as do grey mullet (Mugilidae), some of which feed, at least in part, on detritus.

Phytoplankton has to bloom well if it is to be available in sufficient quantity to sustain microphagous fishes. In the African lakes, species of *Tilapia* that crop phytoplankton are also able to turn to sedimented material of planktonic origin. Their fine food is trapped by mucus produced by glands in the mouth, or filtered through rows of minute spines (microbranchiospines) on the gill rakers (Fryer and Iles, 1972). Zooplankton rarely occurs in sufficient density to permit this kind of filter feeding; instead, zooplankton feeders usually feed like larval fishes, snapping up individual zooplankters. Zaret (1980) observed, for example, that a lake whitefish (*Coregonus clupeaformis*) caught more than 600 *Daphnia* in 15 minutes by 'nipping' them individually from the water! Similarly, deep-sea tripod fish (Figure 2.6) pick copepods from the benthopelagic plankton, as do the smaller damsel fishes (e.g. *Dascyllus* spp.) on coral reefs, and planktivorous clupeids (e.g. *Stolothrissa*) and atherinids in fresh water.

However, there are a few fishes which are large enough to have the gape and gill arch filtering system needed to filter zooplankton. In fresh water, the paddlefish *Polyodon spathula* (Figure 1.11) of the Mississippi basin (up to 1.5 m long) cruises around with its mouth open, filtering cladocerans. In the oceans, examples are whale and basking sharks, and devil rays. Whale-sharks have large gill arches linked by spongy tissue derived from modified denticles. This makes a sieve which enables them to filter small fishes as well as zooplankton (Bigelow and Schroeder, 1948). Sometimes whale-sharks filter in the vertical position, pushing their heads slowly out of the water, allowing the water in the mouth and pharynx to pour out, and then subsiding slowly into the sea again. Basking sharks (10 m or more long) filter copepods with rakers formed from modified denticles. The gill rakers are shed each autumn and a new set develops in the spring, so that the fish cannot feed during the winter. It has even been suggested that basking sharks hibernate resting quietly on the bottom, but raker-less sharks have been caught several times around Plymouth in the winter as they blundered into nets, so this does not seem probable. However, it is clear that basking sharks, living in temperate seas where plankton varies widely in density at different times of year, may balance on the knife-edge of feeding at a loss for part of the year, expending more energy swimming along with widely-opened mouth than they obtain from the plankton they collect. Interesting calculations by Parker and Boeseman (1954) show how delicate the balance must be. Devil rays (Mobulidae) have a single series of gill plates set at right angles to each of the gill arches; they feed on small schooling fish as well as planktonic crustaceans.

7.2.2 Herbivorous fishes

Hickling (1961) pointed out that plant-feeding fishes are in the minority in the sea and in fresh waters of the temperate regions, but are important in tropical fresh waters. Here, seasonal flooding inundates forests and plains and herbivorous fishes can feed on grasses, decaying vegetation, and leaf debris. In forest regions also, plant material falls into the water, providing a continuous source of food for herbivores. Inger and Kong (1962) found that the stomachs of such cyprinids as *Puntius* in North Borneo contained flowers, fruits, seeds, leaves and pieces of stem of vascular plants; they reported that when a fruit or flower fell to the water it was often seized by a fish, usually *Puntius bulu.* Some herbivores feed on attached plants, like the grass carp *Ctenopharyngodon* and various characoids, whilst others, like some cichlids and the South American armoured catfishes (Loricariidae) scrape algae from rocks and stones with their broad-lipped mouth under the snout.

The marine 'forests', where live the greatest diversity of herbivores, are made of calcareous algae and corals. Here certain surgeonfishes (Acanthuridae) rabbitfishes (Siganidae) damsel-fishes (Pomacentridae) and parrotfishes (Scaridae) are among the most important herbivores. Surgeonfishes browse and graze on algae that grow over rocks, corals and sandy stretches of the sea floor. Their closely-set teeth bear a cusped or serrated cutting edge, the entire dentition being thus well suited to the cropping of plants. Incisor-like teeth are also set in the jaws of rabbitfishes. Parrotfishes use their 'beak' to scrape algae from rock and coral surfaces. Certain damsel-fishes, butterflyfishes, blennies and gobies are also largely dependent on algal food, some of which is browsed off hard surfaces. In a recent study of a small Caribbean blenny (*Ophioblennius atlanticus*) which is an algal scraper, Nursall (1981) found that during the day it spent 60% of the time resting, 15% swimming and only 8.5% feeding, which seems little for a herbivore. On the other hand, the intermediate colour phase of an algal feeding parrotfish, *Sparisoma aurofrenatum*, spends 20% of its active time feeding: the larger terminal phase spends 14%.

7.3 Protrusible jaw mechanisms in elasmobranchs

The jaws of most planktivorous fishes (such as clupeids, coregonids, etc.) are non-protrusible, as are those of such freshwater herbivores as characids and catfishes. Marine herbivores are mostly acanthopterygian, members of a group in which the jaws are typically protractile, but their jaws are fixed, as in acanthurids, scarids and siganids. The evolution and retention of protrusible jaws thus seems to be largely correlated with other modes of feeding, and has been of great importance in teleost evolution. The major levels of adaptation in actinopterygian evolution may be defined in terms of jaw morphology, and the radiation of the group has been accompanied by increasing complexity of the jaw apparatus (Lauder and Liem, 1980). Before looking at teleost jaw mechanisms, we begin with elasmobranchs, whose relatively simple mechanisms are certainly designed for a

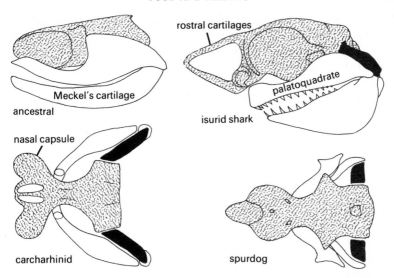

Figure 7.5 Cranium and jaws of different elasmobranchs. Lateral views above, dorsal below. Hyomandibula black; neurocranium and capsules shaded. Note different angles of hyomandibula in different forms. After Daniel (1922), Young (1981), Moss (1977).

carnivorous diet, and which have recently been admirably reviewed by Moss (1977).

In early Palaeozoic sharks, the jaws were more or less terminal, bearing grasping 'cladodont' teeth, and the upper jaw (the palatoquadrate) articulated with the neurocranium through otic and orbital processes, the two palatoquadrates meeting anteriorly in the mid-line. This amphistylic jaw suspension (Figure 7.5) became modified during subsequent evolution as the jaws shortened and the mouth became inferior, to a more mobile hyostylic suspension on the neurocranium. In most modern elasmobranchs, the upper jaw has considerable mobility in the vertical plane and the entire jaw system is braced against the otic region of the neurocranium through the epihyal member of the hyoid arch, the hyomandibular (Figure 7.5).

Feeding mechanisms are best known in the carcharhiniform sharks (e.g. grey sharks). In these forms the hyomandibulae are long and backwardly directed, and the upper jaw, which has prominent preorbital processes, is loosely connected to the front of the brain case by short ethmo-palatine ligaments. The dentition usually consists of awl-like teeth in the lower and broader cutting teeth in the upper jaw. During feeding the hyomandibulae swing outward so as to brace the jaw complex firmly and laterally against the skin behind the mouth. This action also rotates the upper jaw outward and downward and, if the prey is large, facilitates a deep gougelike bite. As Moss (1977) observes, in carcharhiniform and lamniform sharks like the mako and great white, the rotation of the relatively long jaw system and the efficient cutting dentition form a feeding device that is unusual in predators—the ability to shear bites out of prey too large to be entirely encompassed by the open mouth (these sharks also take small fishes and squids).

In squaloid sharks such as the Greenland shark (*Somniosus*), the spur dog *Squalus*, and deep-sea genera such as *Etmopterus*, *Centroscymnus* and *Dalatias*, the hyomandibulae extend outwards at right angles from the brain case to the hinge of the relatively short jaws. The upper jaw is protractile, but the biting mechanism seems not to have the forward rotation that is afforded by the long hyomandibulae of the lamniform and carcharhiniform sharks. In the genera listed above the broader cutting teeth are in the lower jaw. They feed largely on small fishes and invertebrates and according to Moss are able to suck in small benthic prey.

In most rays and some galeomorph sharks, such as orectolobiforms (carpet-sharks, nurse-sharks etc.) and heterodontids (Port Jackson sharks) there is a grasping and/or crushing jaw mechanism bearing setiform holding teeth or molar-like crushing teeth. The hyomandibulae are set in the vertical plane and extend out to short inferior jaws that may carry powerful muscles. In the nurse-sharks (*Ginglymostoma*), cranial muscles raise the hyomandibulae and the palatoquadrates, and antagonistic muscles raise or depress the jaws. The whole system is well fitted to suck in small invertebrates, which are crushed, if need be, before they are swallowed. There are similar sucking mechanisms in diverse rays (e.g. rajids and myliobatids). Indeed, Moss suggests that '. . . the primary importance of hyostylic jaw suspension may have been to allow suction feeding from the benthos'.

7.4 Jaw mechanisms of bony fishes

During the evolution of the ray-finned bony fishes, the jaws became more mobile through changes to three main adaptive levels (Schaeffer and Rosen, 1961). The first innovation occurred during the transitions from chondrostean to holostean forms, when the maxillae were freed from their junctions with the cheek bones and the mandible acquired larger and more complex adductor muscles, coronoid processes, and increased torque around the jaw articulation. The transition from holostean to teleostean levels involved no major changes in jaw mobility but the early teleosts soon evolved a ball and socket joint between the maxillary and palatine bones, which in holosteans are joined simply by connective tissue. In the earliest teleosts and related forms (elopomorphs, clupeomorphs, osteoglosso-morphs, salmoniforms, characoids, etc.) the biting edge of the upper jaw is formed by both the premaxillary and maxillary bones. In more advanced forms (scopeliforms to acanthopterygians) the premaxillae extend back-wards just below the maxillae so as to exclude the latter from the gape. Such exclusion made possible the evolution of protrusible jaws, as seen predominantly in the acanthopterygian teleosts. The main changes involved are the freeing of the premaxillae from the rostral region, coupled with the acquisition of backward and closely adjacent projections (pedicels) that slide over the rostrum; the development of upper jaw ligaments that together with the maxillae control the forward extension of the maxillae; and a forward shift in the suspension of the lower jaw, Figure 7.9 (Schaeffer and Rosen, 1961).

In acanthopterygians, the premaxillary pedicels carry an ethmoid cartilage that slides to and fro over the rostral part of the cranium. Each maxilla hinges near its anterior end (see Figure 7.6) with the maxillary process of the palatine bone and near its posterior end is bound through the lip to the mandible. Thus, as the mouth opens, the backward part of the maxilla is pulled forward and the maxilla rotates about a transverse axis through its palatal hinge. If the pull is exerted on the forward half of the maxilla, as shown by the arrow in Figure 7.6 and which may be through the superficial part of the adductor mandibulae muscle, or by tension on the skin and ligaments if the mouth is widely opened, the maxilla also rotates about its long axis and palatal joint. During this rotation a process on the head of the maxilla presses on an articular surface on the premaxillae causing the latter to slide forwards. Such a mechanism may be found in the perch (*Perca*) and many other acanthopterygian teleosts. But in the cichlid fishes there is no such close articular coupling between the premaxillary and maxillary bones (see Liem, 1980). For a detailed appreciation of all the evolutionary changes involved in the acquisition of jaw protrusibility, especially as studied through electromyography of the cranial muscles and high-speed cinematography, the studies of primitive actinopterygian fishes by Lauder (1980a) may be compared with those by Liem (1980) on cichlids.

Liem (1980) has summarized these and other studies as follows. In advanced teleosts, such as the cichlids and other percoid fishes, mouth opening starts by contractions of the levator operculi and sternohyoideus

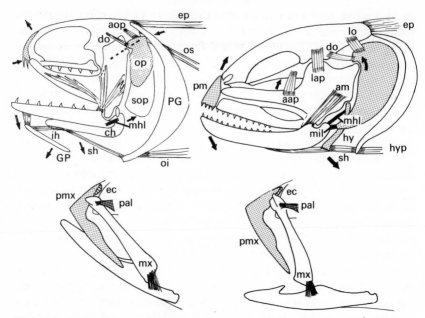

Figure 7.6 Osteichthyan jaw mechanisms. Upper left: reconstruction of muscles involved in jaw opening in a primitive actinopterygian (Paleoniscoid). Dotted line shows position of adductor hyomandibularis; arrows, movements as jaw opens. Upper right: jaw muscles and movements during feeding (arrows) in Arctic char (*Salvelinus fontinalis*). Below: movements of jaws in unspecialized gadoid or percoid, showing the way in which the premaxilla rocks on the cranium via the ethmoid cartilage as the lower jaw is depressed (cf. Figure 7.9). aap, adductor arcus palatini; am, adductor mandibularis; aop, adductor opercularis; ch, ceratohyal; do, dilator opercularis; ec, ethmoid cartilage; ep, epaxial muscles; GP, gular plate; gh, glenohyoideus; hy, hyoid; hyp, hypaxial muscles; ih, interhyoideus; imp, inter-mandibularis posterius; lap, levator arcus palatini; lo, levator operculi; m, maxilla; mhl, mandibulohyoid ligament; mil, mandibulo-interopercular ligament; oi, obliquus inferiorus; op, operculum; os, obliquus superiorus; p. palatine; pm, premaxilla; sh, sternohyoideus; sop, suboperculum. After Lauder (1980), Lauder and Liem (1980), and Alexander (1970).

muscles (see Figure 7.6). The actions of the former muscles depress the mandible by rotating the opercular series of bones backward and upward, thus exerting a backward and upward pull on the articular process of the lower jaw through a ligament between this process and the end of the interopercular bone. A second mouth-opening mechanism soon follows in that the sternohyoideus muscle also depresses the lower jaw by pulling backward and upward on the hyoid arch. This pull is transmitted to the interoperculum through an interoperculohyoid ligament. Epaxial muscles that lift the neurocranium, levator arcus palatini muscles that expand the palate outward, and sternohyoideus, hypaxial and dilator operculi muscles are also active during mouth opening (see Figure 7.6). In effect, the mouth opens rapidly and the buccal cavity suddenly enlarges through the lifting of the head, depression of the buccal floor and lateral expansion of the head. There is thus a sudden reduction of pressure in the buccal cavity that draws in water and prey. Water is expelled when the buccal cavity contracts and the branchiostegal membranes open. In the advanced teleosts, as described earlier, the protractile upper jaw moves forward as the mandible is lowered, which both enlarges the volume of the buccal cavity and directs the suction.

The upper jaw of primitive teleosts, as illustrated by the char (*Salvelinus*) (Figure 7.6) is not protrusible. Moreover, the coupling of the sternohyoideus muscle differs in that its movement is transmitted directly from the hyoid to the mandible through the mandibulohyoid ligament, not through the interoperculum as in advanced teleosts. But earlier in mandibular depression the interopercular coupling comes into play through the action of the levator operculi muscles (see Figure 7.6). As the mouth opens the maxillae swing forward about their forward attachment and the hyoid is more and more depressed. After the mouth is fully open the lateral expansion of the head (through the levator arcus palatini muscles and the dilation of the opercular cavities) reach their peak. As Lauder (1979) has emphasized, the forward swing of the maxillary bones in primitive teleosts brings the buccal chamber forward towards the prey and increases the velocity of the inflow.

Whatever their jaw mechanism, most teleosts generate suction by enlarging their buccal and opercular cavities. Alexander (1970) recorded pressure changes in the buccal cavities of nine species of teleosts ranging from an osteoglossiform (*Papyrocranius*) through a cyprinid catfish and eel to five acanthopterygian species as they sucked in food.

The greatest negative pressure recorded was about 10 kPa for most of the species. As the food was sucked in, the orobranchial chamber expanded by 5–8 cm^3 per 100 g of body weight in most instances (the eels were almost 30 cm long and the other fishes were all between 5 and 12 cm in standard length). More recently, Lauder (1980*b*) has examined suction feeding in the pumpkinseed *Lepomis*, and has proposed a model of the water flow in the mouth cavity during feeding (Figure 7.7) which involves substantially higher negative pressures. Such inertial suction, as Liem (1980) puts it, is the dominant feeding strategy of teleosts. Prey at all levels can be taken effectively by suction. Thus, the cichlid *Petrotilapia tridentiger* (Figure 7.1) which is generally thought to be closely specialized for scraping algae off rocks, was observed by Liem to suck insects and dried fish food from the

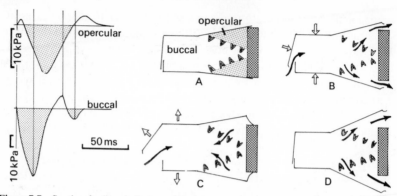

Figure 7.7 Suction feeding. Left: buccal and opercular pressure records from pumpkinseed (*Lepomis*)—note large negative pressures. Right: model showing successive stages in feeding, incorporating backflow from opercular chamber (stippled in A). After Lauder (1980).

water surface, to inhale planktonic prey (by opening and greatly protruding the jaws, then rapidly closing the still protruded jaws) and also to vacuum loose prey (*Tubifex*, midge-fly larvae, earthworms, etc.) from the bottom. All such modes of feeding were by low-speed inertial suction, but through high-speed suction, *Petrotilapia* of 12 to 18 cm in total length fed readily on fish fry up to 6 cm in length. Like other cichlids (e.g. *Petrochromis* and *Hemitilapia*) with the most versatile trophic means, *Petrotilapia* exhibited 8 modes of feeding, as we have already seen.

Not all piscivorous predators inhale their prey by suction. For example, long-jawed fish like *Lepisosteus* catch fish with a sideways strike of the head, and others, such as *Luciocephalus*, open their mouths and then strike forwards to envelop their prey (Figure 7.8) without suction. As Webb and Skadsen (1980) have shown, tiger musky (*Esox*) strike at minnows either by a direct forward movement, or by first bending the body (Figure 7.8) in either case aiming at the centre of mass of the minnow. As over 90% of coral reef fishes are acanthopterygians, and as reefs are trophically diverse habitats for fishes, the extensive adaptive radiation of their protrusile jaw mechanisms is not surprising. Such developments are largely confined to species that take animal food, and these are much the most diverse. In a survey of the food habits of reef fishes in the West Indies, Randall (1967) analysed the stomach contents of 5526 individuals representing 212 species. About 15% of these species were classed as plant and detritus feeders: the rest fed on zooplankton, sessile animals, 'shelled' invertebrates, ectoparasites and fish or were generalized carnivores.

The zooplankton feeders with protrusible jaws take their prey by a

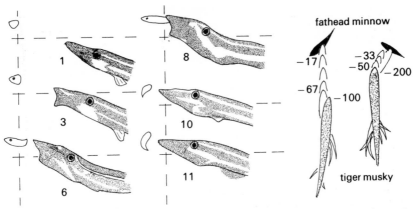

Figure 7.8 Predator strike patterns. Left: frames (numbers) taken at 5 ms apart intervals from film of *Luciocephalus* unsuccessfully attacking a guppy (*Poecilia*). Note cranial flexure and widely-opening mouth as fish attacks; no suction is involved. Right: two attack patterns of *Esox*; numbers represent ms before prey contact. Note that musky aims for centre of mass of prey (fathead minnow). After Lauder and Liem (1981), and Webb and Skadsen (1980).

sudden forward thrust and withdrawal of the jaws. Some of these forms, such as damselfishes (*Dascyllus* and *Chromis* species), and certain butterfly-fishes, feed by day. Forms with larger mouths, e.g. the cardinalfishes (Apogonidae) and squirrelfishes (*Myripristis*), are nocturnal feeders (Hobson, 1974). As Hobson and Chess (1976) have stressed, larger zooplankters including polychaetes, mysids and various peracarid crustaceans, tend to enter the water column by night. During the day some of the fishes that feed on larger prey (e.g. fishes and shrimps) are cryptic ambushing forms like lizardfishes (synodontids), scorpionfishes (scorpaenids) and flounders (bothids); or elongate stalkers like trumpet-fishes (*Aulostomus*), cornetfishes (*Fistularia*), and barracudas (sphyraenids). All such stalkers are elongated in form (Hobson, 1974).

Like the scorpionfishes, other kinds of ambushers (sit-and-wait predators), have large heads and jaws and a relatively small trunk and tail—examples are cottids (sculpins), notothenioid fishes, toadfishes and anglerfishes (Lophiidae). In the Pacific staghorn sculpin (*Leptocottus armatus*), the gape of the jaws is equal to almost one-fifth of the standard length: in a specialized zooplankton feeder, the sockeye salmon (*Oncorhynchus nerka*), the gape is closer to one-twentieth of the standard length (Hyatt, 1979).

Nocturnal and crepuscular habits prevail among the more generalized coral-reef fishes. Some of the advanced forms (e.g. certain butterflyfishes, wrasses and damselfishes) are specialized to extract small prey such as amphipod crustaceans from cover in the benthos and vegetation. Fishes that prey on sessile invertebrates, notably sponges and corals, include other species of the above three families, certain triggerfishes (balistids) and filefishes (monacanthids). There are also species that feed on mobile invertebrates in the benthos. These include the wrasses (Labridae) that pick up molluscs and crustaceans and then crush them between powerful pharyngeal teeth before they are swallowed; red mullet (Mullidae) that detect their prey by mobile chin barbels and often thrust their snout into the sediment in search of polychaete worms and small crustaceans; and various snappers (lutjanids). Besides stalkers and ambushers, fish-eaters include certain moray-eels, groupers (*Epinephelus* spp.), jacks (carangids) and kingfishes (*Scomberomorus* spp.). In the latter forms and their oceanic relatives the tunas and billfishes, the jaws have become non-protrusile.

The wrasses, which are dominant forms in coral reefs and represented also in temperate waters, are among the most advanced of all acanthopterygians. Indeed, the structure and function of their feeding mechanism is very like that in another advanced family, the Cichlidae (see Figure 7.10). Using electromyography, cineradiography and high-speed cinematography, Liem and Greenwood (1981) have observed the mechanisms underlying the biting actions of the upper and lower pharyngeal teeth in such pharyngognathous teleosts. In the cichlids and other

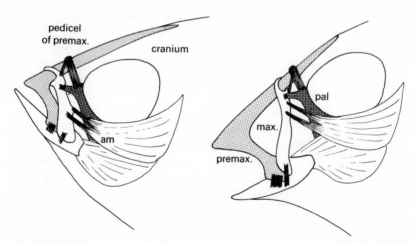

Figure 7.9 The advanced protrusible jaw mechanism of the mojarra (*Gerres*). The premaxillary pedical slides under the palatine ligament to protrude the mouth; the maxilla (max) rotates as the mouth protrudes. After Schaeffer and Rosen (1961).

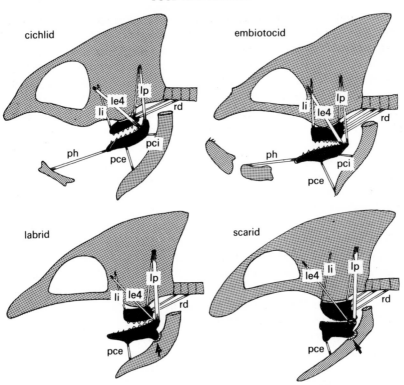

Figure 7.10 Diagrammatic figures of pharyngeal jaw apparatus in different pharyngognath teleosts (pharyngeal jaws black). The upper part of the cleithrum is removed. Note that in wrasses (labrids) and parrotfishes (scarids) the lower pharyngeal jaw articulates with the cleithrum giving better mechanical advantage for a more powerful bite. le4, fourth levator; li, levator internus; lp, levator posteriori; pce, pharyngo-cleithralis externus; pci, pharyngo-cleithralis internus; ph, pharyngo-hyoideus; rd, retractor dorsalis. After Liem and Greenwood (1981).

closely similar forms, which beside the wrasses include parrotfishes and surf-perches (embiotocids), the essential features are that the fourth levator externus and the levator posterior muscles insert on a muscular process of the lower pharyngeal jaw, which is composed of the united 5th ceratobranchial bones. The pharyngeal bite occurs when the backward ends of both the upper and lower pharyngeal jaws move forward and upward towards each other through the co-ordinated action of associated muscles during the masticatory cycle (Figure 7.10). From their structural and functional analysis, Liem and Greenwood (1981) consider that the pharyngognathous mechanisms in cichlids, wrasses, parrotfishes, surfperches and odacids are so closely similar that all these forms may well belong to a monophyletic assemblage.

Beside the cichlids, the cyprinoid fishes are the other major group of freshwater fishes with protrusible jaws and mobile pharyngeal teeth. The jaw mechanism of cyprinids is described by Alexander (1970). Though the jaws are toothless, in diverse species the jaw and pharyngeal mechanisms have become modified to secure and deal with all kinds of food at all levels in the water. There is so much benthic food in both fresh waters and the ocean that fishes with protrusible jaws, both teleost and elasmobranch, almost had to evolve in order to exploit this resource fully. Near the bottom of the deep ocean the most successful fishes in numbers of individuals and species are the rat-tails (Macrouridae). Nearly all macrourids, which are related to cod-fishes (Gadidae), have protrusible jaws much like those of acanthopterygian teleosts. Moreover, most species feed on invertebrate and fish food near and on the deep-sea floor.

The regime of fishes, as will be evident, ranges from omnivorous to more restricted diets. Even so, Liem's (1980a) studies of cichlids has shown that species with a dentition seemingly closely correlated with a special diet are by no means restricted to this diet. Parasitic lampreys, which depend on the blood of other fishes, are presumably the most specialized of all species. The wrasse *Labroides phthirophagus*, for instance, may also be parasitic in that it feeds, if need be, on the scales, skin tissues and mucus of host fishes, but such fishes are not necessarily dependent on the ectoparasites and diseased tissues of their symbionts (Losey, 1972). The very individual and divergent design of the luminous lures of female ceratioid anglerfishes might suggest that each species attracts special kinds of prey. But analysis of the stomach contents of *Oneirodes* species by Pietsch (1974) gave no indication of prey selection. Antennariid anglerfishes (frog-fishes) also have special lures. For instance, *Phrynelox scaber*, which has a worm-like lure, is reported to catch worm-eating fish (Wickler, 1968). But numerous kinds of small fishes eat worms. Even the extraordinarily fish-like bait of an undescribed species of *Antennarius* (Pietsch and Grobecker, 1978) is unlikely to be narrowly attractive.

Close study over several years of fishes that live together in a restricted environment clearly reveals individual differences in diet. This is well shown by Keast's (1970) investigation of food specialization in the fish faunas of some small Ontario waterways. In his Birch Bay study of 15 common species, which included 3 cyprinids, 2 bullheads (*Ictalurus*) and 5 centrarchids, 9 species differed fundamentally in their diets and showed a considerable degree of ecological exclusion. Generalized feeders, such as the blue-gill sunfish (*Lepomis macrochirus*), took all kinds of prey from cladocerans to fish, while the specialists were confined to 2 or 3 food types. The blue-gill feeds at all water levels, whereas a specialist, the filter-feeding *Pomoxis nigromaculatus* (the black crappie) feeds largely on the larvae of a chironomid (*Procladius*) at midwater levels, when these larvae are available. All fish species showed dietary changes from month to month. Indeed, peaks in the numbers of prey species in the environment and in the stomachs of the fishes broadly correspond for Cladocera, Ostracoda, chironomid larvae and pupae, trichopteran larvae and molluscs. There was no such correlation for copepods, which are of little importance as fish food in the area studied.

CHAPTER EIGHT

REPRODUCTION AND LIFE HISTORIES

8.1 Types of life-history

Different fishes produce very different numbers of eggs, from the few of *Latimeria* and elasmobranchiomorphs, to the many millions drifting away in the sea from the spawning grounds of some teleosts. But however fecund, each species must continually and more or less exactly replenish its stocks of breeding adults to compensate for mortality. This chapter compares and contrasts the life-history features of the main fish groups from an ecological point of view. The role of endocrines will not be considered,* but it is important to bear in mind that the interaction between environmental variables such as light and temperature, and the hormones of the pituitary and gonads control fish reproductive cycles, not only by their effects on gametogenesis but also in sexual differentiation, sexual behaviour and fertilization.

Fishes have two main kinds of life histories. In the first, as in lampreys and most bony fishes, many small eggs are produced. After fertilization, these develop into larvae which feed and grow until they are ready to metamorphose into young looking more or less like their parents. In the second, as in hagfish and elasmobranchiomorphs, a few large yolky eggs develop directly into young of adult form. Apart from the handful of rays living in fresh water (Chapter 6), the large-egg producers are all marine. Small-egg producers are found in both marine and fresh water (Figure 8.1).

8.2 *K*- and *r*-selection

The contrast between these two types of life-history has been expressed in terms of the selection pressures operating in different environments; the *K*-

*The reader is referred to *Fish Reproduction: Behaviour and Endocrinology* (a symposium in honour of W. S. Hoar), *Can. J. Fish. Aq. Sci.* **39** (1982) 3–137.

148

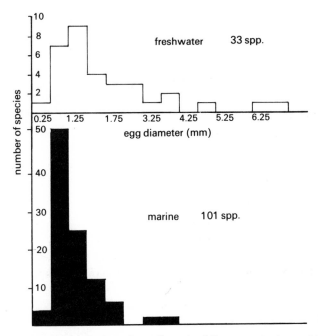

Figure 8.1   Frequency distribution of egg diameter of European marine and freshwater fishes. After Woolton (1979).

selected species of theoretical ecologists live in stable, crowded environments, whilst r-selected species are fitted for maximal population growth in uncrowded, but less stable, environments. The concepts of K- and r-selected species were derived by MacArthur and Wilson (1967) from the logistic equation for population growth, expressed simply as

$$\frac{dN}{dt} = \frac{rN(K-N)}{K}$$

(where N, number of individuals present; r, rate of increase; t, time; and K, an upper asymptote). The equation expresses the growth of a population increasing steadily in numbers to an upper asymptote K. The term $(K-N)/K$ expresses the numerical separation of the population from the asymptotic level of growth, and the carrying capacity of the environment is related to K, which MacArthur and Wilson saw as referring to selection for competitive ability in crowded populations. In contrast, r is an expression of selection for high population growth in uncrowded environments.

These concepts have been expanded to cover other life-history traits:

Stearns (1976) lists K-selection as favouring slow development, delayed reproduction, large body size, and low resource thresholds, whilst r-selection favours rapid development, early reproduction, small body size, high resource thresholds and maximum rates of population increase. The significance of these two very different kinds of selective pressures has recently been discussed by Parry (1981), and Gould (1977) has given a stimulating (if not entirely convincing) consideration of their roles in the origin of new animal groups. Although the life-histories of most fishes can fruitfully be considered in terms of K- and r-selection, there are some difficulties, and as the philosopher J. L. Austin sagaciously remarked in another connection '. . . it is essential, here as elsewhere, to abandon old habits of *Gleichschaltung*, the deeply ingrained worship of tidy-looking dichotomies'.

8.3 Agnatha

Both lampreys and hagfishes have an unpaired gonad without a gonoduct. The eggs or sperm are shed into the body cavity and then extruded through abdominal pores into the water. In the large sea lamprey (*Petromyzon marinus*), egg counts have ranged between 24 000–236 000, whilst in smaller non-parasitic forms there may be 400–9000 eggs. The eggs are laid in 'nests', or redds, made in the stream bed, and hatch in a fortnight or so as small pro-ammocoete larvae which are in many respects 'prototype' chordates (Whiting, 1972). These soon change into the active ammocoete stages which burrow into silt banks and filter-feed for several years. After five or more years, the ammocoetes metamorphose into adults (Hardisty, 1979). Generally, larval life is longer in parasitic species than in non-parasitic (Figure 8.2).

Annual mortality of the ammocoete stages is relatively low, judging from the decrease in population size of the year classes of ammocoetes, and probably the main mortality is during hatching. Tests under natural conditions showed that only 5.3–7.8 % of all the eggs produced by a sea lamprey develop into viable ammocoetes (Hardisty and Potter, 1971). Adult existence may also be hazardous for a parasitic species, especially when host fishes of suitable size are scarce, and Hardisty (1979) points out that lack of hosts might favour prolongation of the stable protected larval phase and delay metamorphosis, leading to smaller non-parasitic forms. However caused, the repeated evolution of non-parasitic lampreys from parasitic forms is unique amongst vertebrates.

No other fish larva leads the long sheltered existence of an ammocoete

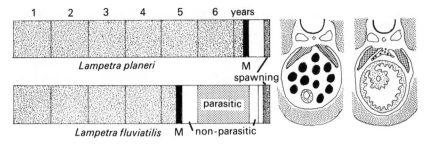

Figure 8.2 Diagrammatic illustrations of life history of parasitic and non-parasitic lampreys. Right: cross-sections of body of parasitic and non-parasitic (left) lampreys after metamorphosis showing few large oocytes in non-parasitic and many small oocytes in parasitic forms. After Hardisty (1979).

(or has such a microphagous means of subsistence). Whilst in the larval stage lampreys are evidently like K-selected organisms in their size and slow maturation; however, their food source is not limited, and in this, and the high fecundity of the adults, lampreys are more like typical r-selected species.

Hagfishes fit better as K-selected organisms, for they lay few large ellipsoidal eggs which hatch after 2 months or more as small versions of the adult. The eggs are laid from time to time on the sea bed linked in batches of 5 or 6, and are some 14–25 mm long. Curiously enough, although those of *Bdellostoma* are well known, the eggs of the European *Myxine* are very rare. Holmgren (1947) found only 131 during a 20-year search, and almost all of these were undeveloped. The prize offered in 1865 by the Copenhagen Academy of Sciences for a description of the embryology of *Myxine* remains unclaimed! Hagfishes are probably functionally dioecious, although their gonads pass through an hermaphroditic phase (Walvig, 1963); as in elasmobranchs, the disadvantages of low fecundity are evidently counterbalanced by the production of young of adult form at hatching.

In their few large eggs, slow development and life in sizeable, crowded local populations, hagfish are typical of K-selected species; the tendency of hagfishes to be deep-sea forms accords with Cody's (1966) view that the degree of K-selection is correlated with the stability of the environment.

8.4 Elasmobranchiomorpha and *Latimeria*

Elasmobranchs and holocephalans have paired penes or claspers formed from the inner elements of the pelvic fins that roll up to form a tube with

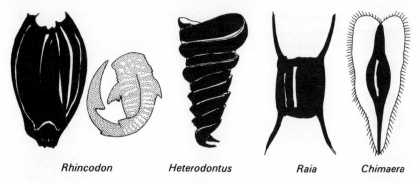

| Rhincodon | Heterodontus | Raia | Chimaera |

Figure 8.3   Egg cases of oviparous elasmobranchiomorphs (not to same scale). After Lineaweaver and Backus (1970), Dean (1906), Daniel (1922).

overlapping edges. These male claspers transfer sperm to the female to fertilize the eggs in the oviduct, after which they may either be retained, or laid on the sea bed. In oviparous species, each egg with its massive yolk store enters the shell gland of the oviduct (where spermatozoa are stored) and when fertilized is there encased in a horny capsule. Oviparity is confined to the chimaeras (Holocephali); the families of skates (Rajoidea); and four families of sharks, Heterodontidae (Port Jackson sharks), Rhincodontidae (whale-shark), Orectolobidae (carpet-sharks etc., certain species are viviparous) and Scyliorhinidae (dogfishes; one species is viviparous). Eggs range in size from over a centimetre in dogfishes to the 30×15 cm egg case of a whale-shark (Figure 8.3); the eggshell of the chimaeroid *Callorhynchus* measures about 25 cm in length. Recorded incubation times are $2\frac{1}{2}$–3 months in a carpet-shark (*Chiloscyllium griseus*), $4\frac{1}{2}$–8 months in *Raja* species, 6–8 months in *Scyliorhinus* and 9–12 months in the Port Jackson shark (*Heterodontus*) and the chimaera *Hydrolagus colliei*.

Viviparous species have either placental or aplacental (ovoviviparous) means of development. The placental form is confined to the hammerhead sharks (Sphyrnidae) and the most diverse of all shark families, the Carcharhinidae (grey sharks). In the smooth dogfish *Mustelus*, a member of the latter family, the embryo is dependent on its yolk for about 3 months (out of a gestation period of 10–11 months) and thereafter on a placenta. Each of the embryos develops in a separate maternal compartment in which villi are formed that fuse with grooves in the yolk sac. In species of the small shark genus *Scoliodon*, and in hammerheads, there is a closer union between the yolk sac of each embryo and the maternal epithelium.

The maternal mucosa forms a series of small highly vascular cups and to each one a yolk sac is joined. Moreover, a placental cord is formed bearing thread-like processes that aid in the absorption of the maternal secretions (Figure 8.4).

The remaining viviparous sharks are aplacental, and so are the non-oviparous rays. In the latter the yolk sac is rudimentary, but placental analogues in the form of elaborate glands in the oviduct secrete an albuminous uterine milk, which in some species is aspirated into the oral cavity of the embryo through its spiracles. In species such as the sting-ray *Dasyatis violacea*, the eagle-ray, *Myliobatis bovina*, and the butterfly-ray *Gymnura micrura*, the gestatory parts of the oviduct form thousands of long nutritive threads (trophonemata). These enter the embryo through the spiracles and extend into the oesophagus, where their secretory products are released (Figure 8.4).

The embryos of aplacental (ovoviviparous) sharks either depend on the yolk reserves or are egg-eaters. In the spiny dogfishes (Squalidae) there is an internal as well as an external yolk sac. The internal one is developed from an expansion of the yolk stalk and it extends into the body cavity of the embryo, where it opens into the intestine. Ciliated epithelia in the

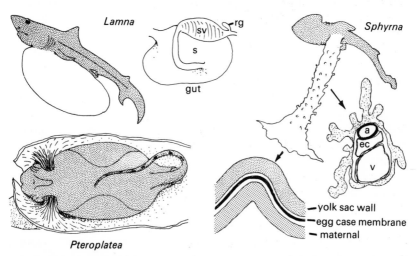

Figure 8.4 Elasmobranch embryonic nutrition. Upper left: egg-eating embryo of porbeagle (*Lamna*) showing greatly enlarged stomach (seen in detail on right). Lower left: embryo of butterfly ray (*Pteroplatea*) showing trophonemata entering spiracles. Right: placenta of the hammerhead (*Sphyrna*). The embryo is attached to the placenta by a cord bearing absorptive processes, seen in section to the right. a, artery; v, vein; ec, extra-embryonic coelom; rg, rectal gland; s, stomach; sv, spiral valve. After Alcock (1892), Gilbert (1958) and Shann (1911).

external yolk sac transport yolk through the internal yolk sac and thus into the intestine, where it is digested and the products absorbed. In *Squalus acanthias* this nutritive system is formed and functional when the embryos are 60–70 mm in length. During the earlier part of gestation, nutrition comes directly from the external yolk sac, which disappears before parturition (Hoar, 1969). Oophagy has been studied in the mackerel shark, *Lamna*, and the sand shark, *Odontaspis*. In the former the yolk is soon absorbed, but immature eggs that pass into the oviduct and thence to the uterine chambers are eaten by the embryos (Figure 8.4). After a gestation period of 2 years (Holden, 1974), the young (1–4) have reached a length of about 60 centimetres. Presumably the embryos of the basking shark, *Cetorhinus maximus*, which produces many small eggs, are also oophagous.

Part of the foregoing review of reproduction and development in cartilaginous fishes is based on Wourms (1977), who also considers the factors relating oviparity and viviparity. In phylogenetic perspective, oviparity is the least specialized mode of reproduction, and from such a beginning viviparity was evolved independently in the different major groups. If geographical factors are at all significant, they are so only in the rays (Batoidei), whose oviparous species (rajoids) live mainly from temperate to polar regions: the viviparous sawfishes, guitar-fishes and rays are centred in subtropical and tropical regions. Oviparous sharks and rays tend to be bottom-dwellers of inshore waters and of relatively small size. The larger sharks and rays are viviparous and they also produce larger young (mostly 30 to 70 cm in length) than do oviparous species (< 30 cm). Even the whale shark, the largest of all sharks and oviparous, must produce much smaller young (in a 30×15 cm egg case) than produced by the largest viviparous species, the basking shark, whose smallest known young are 150 to 180 cm long.

Compared to the smaller young of oviparous species, those of viviparous species should not only have fewer predators and less competition, but also better swimming powers to take advantage of a greater potential range of food organisms. Moreover, during the gestation period the developing young are protected from predators and live in physiologically regulated surroundings. If we consider the mackerel shark (*Lamna nasus*) and the thresher shark (*Alopias vulpes*), which have a probable gestation period of two years (Holden, 1974) at the end of which there are no more than 5 young, it would seem that one disadvantage of viviparity is a much reduced rate of reproduction. On the other hand, the blue shark (*Prionace glauca*) has a maximum litter size of 54 young in Atlantic waters, but their

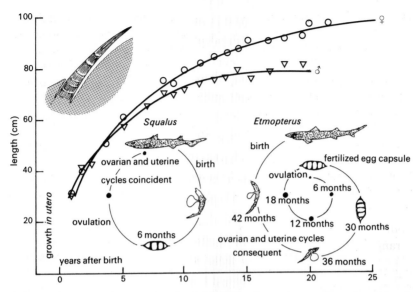

Figure 8.5 Growth of the spur dog, *Squalus*. Calculated growth curves for male and female spur dogs, with observed points. Below, reproductive cycles of *Squalus* and *Etmopterus*. Inset: dorsal fin spine showing pigmentation. After Holden (1974) and Holden and Meadows (1962).

mean weight is only 0.14 kg and they measure 31 to 47 cm. The young of the mackerel shark weigh about 10 kg and are about 70 cm in length. However, as adults of the largest (and viviparous) kinds of elasmobranchs have few predators, it may well be, as Holden (1974) suggests, that the mortality rate of those species producing large numbers of young is higher in the early years of life than in those species producing small litters, especially as the size of the young is inversely related to litter size.

To study growth rates one needs some means of determining age. In well-studied species, such as the spiny dogfish, *Squalus acanthias* and the skate *Raja clavata*, annual rings seen in transverse sections of the spines and in the vertebral centra (of the second) have been found in North Sea stocks (Figure 8.5) (Holden, 1974). Where such growth rings may not be evident, as in subtropical and tropical waters, one would hope to turn to studies of year classes for annual statistics, but very few elasmobranchs have been commercially exploited. As data are thus limited, Holden has used an equation derived from a modification of the von Bertalanffy equation, which is feasible because advanced young result from a long period of hatching or gestation. The assumption is that growth curves of the embryos can be extrapolated to give growth curves of free-living individuals. The derived equation is

$$l_{t+T}/L^{\infty} = 1 - \exp(-KT)$$

where l_{t+T} is the length at birth, L^{∞} the maximum observed length (L_{max}), T the length of gestation or hatching period, and K a constant.

For instance, the maximum observed length of *Squalus acanthias* in European waters is 108 cm, the length at birth 27.5 cm and the period of gestation 2 years. Thus $27.5/108 = 1 - \exp(-2K)$, giving $0.7454 = \exp(-2K)$, $2K = 0.294$ and $K = 0.15$. The derived growth constant K may be compared with the value 0.11 obtained from growth data. The fit between values of K for female fish calculated from growth data, compared to those estimated from ratio of length at birth to the maximum observed length, is exact in the basking shark (0.12) and the soupfin shark (*Galeorhinus australis*) (0.09) and quite close in species of skates (*Raja*) (0.13–0.31, cf. 0.20–0.46). In general, values of K seem to be 0.1–0.2 for sharks and 0.2–0.3 for rays (Holden, 1974). But if the growth constant of the spiny dogfish is typical of sharks, then 7 species of smooth hounds (*Mustelus*) are exceptional. Their gestation periods range from 10–13 months, sizes at birth from 28–43 cm and lengths at maturity from 60–100 cm. When Francis (1981) put the values for each species into the modified von Bertalanffy growth equation, he obtained growth constants K ranging from 0.22 to 0.53 for males and from 0.21 to 0.36 for females. Moreover, sexual maturity is attained 1–4 years after birth, which indicates that smooth hounds grow a good deal quicker than spiny dogfishes—it takes 10 years for *Squalus acanthias* to mature (Holden, 1974).

Oviparous rays are not only among the quicker-growing forms, but are also the most fecund of the elasmobranchs. Observations on skates (*Raja* spp.) suggest that the number of egg capsules produced by a female during a year is usually somewhat less than 100. (Some of the ovoviviparous species only produce 1–4 young per year). Thus, the fishery for skates in the North Sea, Irish Sea and Bristol Channel showed that the catch rates remained quite stable for 20 years, but they have declined during the past few years, when the fishing effort has eventually reduced the rate of recruitment to the stocks. Indeed, as Holden (1974) argues, all the available data indicate that fisheries for elasmobranchs are not sustainable. As we have seen, elasmobranchs have low rates of reproduction, they grow slowly and most species take a long time to mature (calculations suggest 7–13 years in sharks at 50% maturity and 5–6 years in rays). In ecological terms they are K-selected species, and so, we may

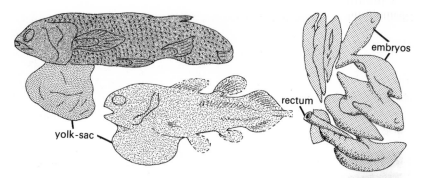

Figure 8.6 Coelacanth embryos. Above: *Latimeria* embryo from oviduct, shown schematically on right. Below: *Rhabdoderma* (Upper Carboniferous)—note yolk sacs in fossil and recent embryos. After Smith *et al.* (1975) and Schultz (1977).

presume, is the (ovoviviparous) coelacanth, *Latimeria*, which incubates very large eggs over a long period and lives to an age of 11 years or more (Hureau and Ozouf, 1977). *Latimeria* embryos have a large yolk sac, seen also in some fossil coelacanths (Figure 8.6).

8.5 Teleosts

Teleost life histories are various, but in only two major groups above family level (ophidioids and cyprinodontoids) are viviparous forms well represented. The majority lay eggs which develop into small larvae which have a relatively short larval life of high mortality. The three-year drift across the Atlantic of larval European eels (*Anguilla anguilla*) is exceptional. Perhaps the transparency, serpentine movements and large size of leptocephali (Figure 8.7) are good defences against predators (Marshall, 1979).

8.5.1 Freshwater species

The killifishes, known to aquarists by more than 150 species, are small, fresh- and brackish-water fishes from tropical to temperate regions of the world. The viviparous killifishes (Poeciliidae) and the oviparous killifishes (Cyprinodontidae) are the most diverse families. In the former the male has a mobile intromittent organ, the gonopodium, formed from elongated rays of the anal fin, and the female is usually the larger sex; the male is generally the larger in oviparous killifishes. Certain viviparous species are gynogenetic, all-female forms. For instance, out of 18 species of *Poeciliopsis*, which

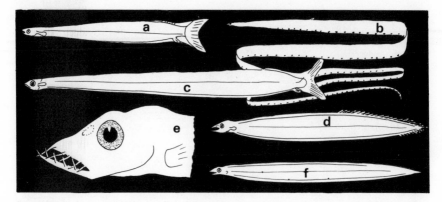

Figure 8.7   Leptocephalus larvae. (*a*) *Elops*; (*b*) notacanth (perhaps *Aldrovandia*, Figure 2.5—this larva can almost certainly exceed a metre; (*c*) *Albula*; (*d*) *Platuronides* (Serrovomeridae); (*e*) head of *Nemichthys* (snipe eel); (*f*) *Gnathophis* (Congridae). Not to same scale. After Alexander (1961), Beebe and Crane (1937), Castle (1963, 1959), Smith (1970) and Hildebrand (1963).

live primarily along the Pacific coast of Mexico, six are unisexual. The eggs of the latter require sperm from a related species to stimulate their development (Schultz, 1977). One oviparous species, *Rivulus marmoratus*, is a self-fertilizing hermaphrodite (Harrington, 1971).

The eggs of oviparous killifishes are large (Rosen, 1973) and bear adhesive filaments. They may be laid on the bottom, hidden, buried or carried about, and hatch into miniature adults that are ready to swim and feed. The same is true of the newly-born individuals of viviparous species.

Most of the egg-laying killifishes belong to the subfamily Rivulinae, nearly all of which are less than 7.5 to 10 cm in length and live for 3 or 4 years. But in certain genera there are 'annual' species with a life-history fitted to exploit the temporary pools that appear each year in tropical and subtropical regions, as in South America and Africa (Simpson, 1979). These pools form during the rainy season, and during the several months that they may last before drying up, produce abundant small-fish food, such as copepods, ostracods and mosquito larvae. Annual killifishes (such as species of *Aphyosemion* and *Nothobranchius* in Africa and species of *Austrofundulus*, *Cynolebias* and *Pterolebias* in South America), survive the dry season by producing eggs made drought-resistant by the development of thick chorionic membranes. These eggs are buried in the mud, and after an embryonic period that may last from a few weeks to several months, they hatch quickly when the rains return (Figure 8.8). During the

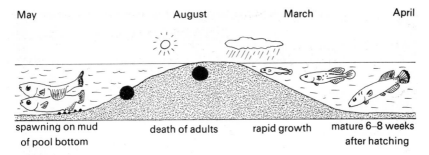

Figure 8.8   Life cycle of annual killifishes. After Simpson (1979).

embryonic period the extraordinary adaptive feature is an insect-like suspension of growth and development in a diapause phase, which may occur at two or three different stages of embryogenesis. In many populations there are also 'escape' eggs that avoid diapause at a particular phase. The young grow rapidly after hatching and in six to eight weeks are ready to spawn. During their reproductive life the daily fecundity of annual killifishes, depending on the species, size, food supply and age, seems to range from a few eggs to at least 50 (Simpson, 1979).

Annual killifishes have a weed-like capacity to flourish in temporary habitats and survive adverse conditions. In their smallness, rapid development, early reproduction, and short life they could be classed as r-selected species. But unlike typical r-selected forms, they are iteroparous rather than semelparous, and their reproductive effort, considering their size and egg size (diameter ca. 1.0–3.0 mm) is modest. Moreover, in killifishes at least, the opportunistic species are not necessarily those that lay eggs. The mosquito-fish, *Gambusia affinis*, so called from its great liking for the larvae and pupae of *Anopheles*, is a viviparous killifish (males 3.5 cm; females 6.5–8.0 cm), which produces a brood of 43 to 205 young once in 21 to 28 days (see Miller, 1979). This species not only proved to be better than other small fishes as a means of malarial control, but soon became acclimatized after its introduction to many parts of the world. *Gambusia affinis* occurs naturally in south-eastern parts of the United States, but it is now perhaps the most widely distributed of all freshwater fishes (Krumholz, 1948). If frequent enough, the repeated broods of viviparous killifishes may make them not much less fecund than their oviparous relatives. Indeed, in certain poeciliids the storing of sperm in the ovarian wall and the gestation of the embryos within intact egg follicles, has made possible the development of super-foetation. Through a succession in the ripening and fertilization of the egg batches, a number of broods, each in its own stage of development, can be gestated in the ovary. In *Poeciliopsis retropinna, P. elongatus* and *Heterandria formosa,* up to 9 broods may be present at any one time.

Even so, the relatively low birth rates of viviparous killifishes, compared to those of small oviparous species in other teleost orders, seem particularly risky in environments subject to marked seasonal fluctuation. Perhaps this partly explains why viviparity in killifishes and other small fishes is largely confined to tropical and subtropical forms. In North America the killifishes most at home in temperate conditions belong to the oviparous genus *Fundulus* (see also Miller, 1979).

In wider perspective, the spawning of many freshwater fishes is linked, like that of annual killifishes, to rainy seasons in the tropics. Indeed, in flood-plain rivers the entire life of a fish community is geared to the rains, when flooding greatly increases the size of the aquatic environment, and also adds nutrients which evoke blooms of phytoplankton and concomitant increase in food organisms (Lowe-McConnell, 1975). Many of the larger fishes, especially the ostariophysans, spawn just before or during the floods, while others spawn in grass swamps at the edge of the advancing floods. For many tropical fishes the high-water season is also the main feeding and growing season, when they build up fat stores to help them through the next dry season.

In great tropical lakes conditions are more stable and become more or less non-seasonal. Thus, in the equatorial African Lake George and Lake Victoria, ripe individuals of most kinds of cichlid fishes exist at any time of the year. But there may be seasonal peaks in the numbers of spawning individuals. The fish faunas of the Great Lakes of Africa are dominated by cichlid fishes which have formed species flocks of more than 300 species in Lake Malawi; over 200 in Lake Victoria; and 126 in Lake Tanganyika: nearly all are endemic to the lake in question. For the most part these cichlids produce small egg batches that are mouth-brooded by the female. In the substrate spawners, like certain species of *Tilapia,* both males and females guard the eggs and young.

Lowe-McConnell (1975) ends her review by contrasting the fluctuating fish populations of seasonal tropical waters with those of stable, aseasonal habitats. In the former waters, fish populations fluctuate markedly through migrations, seasonal spawning and mortality. Such conditions involve strong selection pressures for high fecundity, rapid development and growth, short life cycles and a rapid turnover of the population (as shown by a high productivity (P) to biomass (B) ratio. In contrast fish populations of aseasonal habitats are very stable and here selection is for reduced fecundity (as in brooded eggs), longer life cycles and a lower turnover of the population (shown by a low P/B ratio).

Since the body fluids of freshwater fishes have a higher solute content than an equivalent volume of their surroundings (Chapter 6), it is hardly surprising that they lay non-buoyant eggs. Such eggs may be buried, attached to vegetation, placed in a nest, carried, or, as we have seen, brooded (see also Marshall, 1965). Though a few species, such as gouramis and grass carp, do produce buoyant eggs, non-floating kinds are best suited to most freshwater habitats. Land waters eventually flow into the sea but the eggs and larvae of fishes (and other animals) need to be kept in one place if local populations are to be maintained. Moreover, freshwater fishes tend to lay larger eggs than do those marine species that produce buoyant eggs. For instance, in freshwater cyprinids of the temperate Northern Hemisphere, the egg diameter varies between 1.0 and 2.0 mm in the females of nearly all species ranging from about 5 to 60 cm in length (see Miller, 1979). In a comparable series of marine fishes from north temperate waters (see Russell, 1976) the egg diameter of nearly all species producing buoyant spawn is closer to 1.0 than to 2.0 mm. Thus, the newly-hatched larvae of the freshwater species will tend to be larger than those of the marine species. The larger the larvae the more easily will they be able to swim and survive in waters that move toward the sea.

8.5.2 Marine species

Producers of buoyant eggs. Marine teleosts, whose body fluids have a salt content lower than that of the sea, seem to be admirably preadapted to produce buoyant eggs (see Denton, 1963). As each egg nears maturity dilute body fluids are secreted by surrounding follicle cells. More than any other component of low specific gravity, such as oil globules, these follicular fluids make the eggs neutrally or even positively buoyant. And in newly-hatched larvae the same fluids are retained in subcutaneous buoyancy chambers (Figure 8.9).

Thus, buoyant eggs and larvae, and very soon post-larvae with a small gas-filled swimbladder, are dispersed by the sea from the spawning ground. During their drift, the post-larval stages hunt small food organisms, such as copepod nauplii, in the plankton, the most successful growing most rapidly towards the metamorphosis stage. During metamorphosis, trenchant changes in body form and inner organization, but not in body size, turn a post-larva into a young fish with much of an adult appearance.

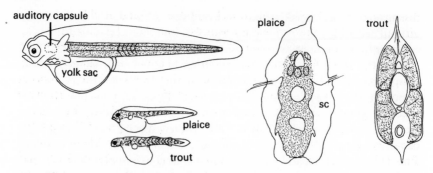

Figure 8.9   Pelagic and non-pelagic teleost larvae. Left: plaice larva; below, plaice and trout larva compared. Right: transverse sections of plaice and trout larvae showing sub-cutaneous buoyancy spaces (sc) in plaice. After Shelbourne (1955).

After metamorphosis the fry begin to move towards the environment of their parents.

This kind of life cycle, characterized above all by a rapidly growing dispersal phase in the plankton, divided by metamorphosis from a growth towards maturity, is shown by many kinds of marine invertebrates and teleost fishes. Two living spaces are involved, separated by metamorphosis. Wilbur (1980) calls this a complex life cycle. On land it occurs in many insects and amphibians.

Out of some 12 000 species of marine teleosts, perhaps as many as 9000 produce buoyant eggs. But as we shall see, in moving from tropical to polar regions there is a decided cline towards the production of relatively large grounded eggs. Beginning with the fishes of coral reefs and atolls, very many species produce large numbers of pelagic larvae that are dispersed in off-lying open waters. Moreover, in most species these larvae hatch from pelagic eggs, as laid for instance by sea-basses (Serranidae), snappers (Lutianidae), red mullet (Mullidae), and butterflyfishes (Chaetodontidae). Gobies, blennies, and certain damselfishes (Pomacentridae) are among the groups that produce non-buoyant eggs, which may be either scattered over the bottom or attached somewhere and guarded by one or both parents. Most species spawn several times during a long breeding season and in all, as in serranids, and surgeonfishes, total fecundities may range from tens of thousands to millions of eggs. Such reproductive activities, as Sale (1980) observes, seem directed towards broadcasting pelagic eggs and larvae for dispersal in the open ocean. During this dispersal mortality must be very high, as shown by the numbers of young returning to reef waters. By circulatory and other unknown means small numbers of recruits are disengaged from the plankton to seek suitable reef and atoll habitats, but virtually nothing is known of the size and extent of local populations.

The eggs of coral fishes take no more than a day or so to hatch and subsequent larval life (about 8 weeks) may be about half that required in temperate seas. There is certainly evidence that fishes from coral-bearing regions have the right reproductive rates and life cycles to be opportunistic species. Since the opening of the Suez Canal in 1869 at least 27 species of Red Sea fishes (and numerous invertebrates) have migrated through the canal and established themselves in parts of the Eastern Mediterranean. Por (1978) who calls them 'Lessepsian migrants' (Figure 8.10), has analysed their spread in seas that must suit them physically (temperatures in the subtropical Levant Basin range from 16°–29°C and salinities reach 39.55°/₀₀) and biologically—the eastern part of the Mediterranean is a relatively deserted area, affording living space to the invaders. Most of the migrators have reached the Levant shore (11.6% of the fish species off Israel are Red Sea migrants), while certain species, such as the squirrelfish (*Holocentrus ruber*) (Figure 8.10) and the rabbitfish (*Siganus rivulatus*), have spread westward along the Anatolian and North African shores. Measured by area colonized, other successful invaders include a round herring (*Dussumiera acuta*), a lizardfish (*Saurida undosquamis*) a barracuda (*Sphyraena chrysotaenia*) and a red mullet (*Upeneus moluccansis*). Such

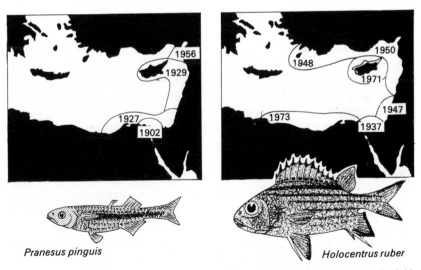

Pranesus pinguis

Holocentrus ruber

Figure 8.10 Two Lessepsian migrants. Distribution in the Mediterranean of the atherinid *Pranesus*, and the squirrelfish *Holocentrus* at different dates. After Por (1978).

fishes, as Por (1978) indicates, may be seen as expansive r-selected forms, and he adds that there is no evidence yet of a reversal in their spread.

In the open ocean of the tropical and subtropical belts, epipelagic teleosts such as flying-fishes (Exocoetidae), sauries (Scomberesocidae), sailfishes and marlins (Istiophoridae), tunas and ocean sunfishes (Molidae), also produce numerous buoyant eggs. So do nearly all of the fishes that live below them in the mesopelagic and bathypelagic zones. Their eggs range in diameter from 0.5 to 1.65 mm. Species of both levels are presumed to spawn where they live, and the eggs float upwards towards the surface. The youngest stages of mesopelagic fishes such as lanternfishes (Figure 1.3), and of bathypelagic fishes like ceratioid anglerfishes, are well known from surface waters. Here the larvae feed and grow until metamorphosis when the young move down towards the adult habitat. During these ontogenetic migrations, which from spawning to descent of the fry must extend over a month or two, mortality is likely to be very high. Adaptations to counter mortality are considered by Marshall (1979).

Both in the open ocean and in coastal waters, species diversity drops markedly on moving from subtropical to temperate latitudes. In temperate coastal waters, as implied already, the species producing pelagic eggs do not markedly outnumber those that produce grounded or other kinds of eggs. Thus, 68 of the 111 species of British marine fishes studied by Russell (1976), lay floating eggs. Of the 43 species that produce grounded eggs, the main forms are gobies, sand-eels (Ammodytidae), blennies and clingfishes (Gobiesocidae). There are also two viviparous species of *Sebastes*; the marine stickleback (*Spinachia*), which lays its eggs in a nest; and six species of pipefishes (Syngnathidae), whose males carry the eggs and young in a brood pouch. As elsewhere in the ocean, buoyant eggs have a crystalline transparency, whereas grounded eggs, which are larger and have tougher membranes, are much more opaque and may be suffused with yellow or orange pigments. Harmful rays at the blue end of the spectrum will pass through a transparent egg floating near the surface, or be absorbed by the membranes and colour filters of grounded eggs (Breder, 1962).

Most of the fishes that lay non-buoyant adhesive eggs live in near-shore waters where waves search and scour the shallows, impelling long-shore and rip currents. Anchored eggs fit best in such surroundings, where buoyant eggs would often be cast ashore, but an attached egg mass makes excellent food. The common habit of guarding the eggs (e.g. in gobies, lumpsuckers, clingfish and toadfish), which may be hidden in elaborate nests (wrasse, sticklebacks) perhaps arose to thwart egg-eaters. Usually the male is the guard, threatening intruders and fanning water over the eggs to keep them free from silt and microorganisms.

Because of its exploitation, or rather over-exploitation, by man, the herring (*Clupea harengus*) is the best known of the fishes that lay grounded eggs. For the same reasons, the most investigated of the species producing buoyant eggs are such fishes as cod (*Gadus morhua*), plaice (*Pleuronectes platessa*) and mackerel (*Scomber scombrus*). Each species is represented by separate stocks, each covering part of a general area of distribution. In temperate waters the centres of these adult stocks are linked to spawning areas and nursery grounds in ways modelled by Harden-Jones' (1968) triangle of fish migration. Spawning grounds, which are circumscribed in place and time, are reached by mature fish swimming against the current. During the breeding season, larval stages drift away from these areas towards nursery areas, usually closer inshore, where they metamorphose and leave the plankton. During subsequent growth the fish spread into deeper water and eventually reach the grounds of their stock. The migrant circuits of plaice in the southern North Sea are described by Cushing (1975). The surveys of fishery biologists also reveal how narrowly the maintenance of stocks may depend on fecundities large enough to cover the inroads of mortality, which is bound to be heaviest during the early life history. Cushing (1975) estimates death rates of eggs and early larvae to be as high as 5–10% per day, though such mortality obviously subsequently declines.

Arctic and antarctic coastal waters are near freezing and covered with ice for most of the year. Virtually all of the resident fishes, so far as we know, lay large yolky eggs on the sea floor (Marshall, 1953). In the notothenioid fishes of antarctic regions, egg diameters range mostly from 2.0 to 5.0 mm and fecundities from 2500 to 120 000. Due to slow growth rates, maturity is not reached until fish are several years old (Everson, 1977). If species spawn during the autumn and hatching takes about 3 months, post-larval stages will appear in spring, the start of the short growing season in the plankton, although some species spawn during the summer (Marshall, 1953; Everson, 1977). At all events large eggs yield large larvae that, judging from Regan's (1916) figures, are soon ready to take a relatively wide size range of food organisms in the plankton or elsewhere. Wyatt (1980) points out that fish in these waters must choose either to adopt a short-lived larva exploiting the short growth season, or to become independent of season by developing large demersal eggs which hatch into larvae that feed on or near the bottom, where seasonal changes are least. Most Arctic fishes have chosen the second alternative; the demersal eggs of eel-pouts (*Lycodes*), lumpsuckers (*Eumicrotremus*) and wolf-fishes (*Anarhichas*) range from 3–8 mm, and their larvae feed on or near the bottom.

Such fish do not need to time their spawning with the plankton peak, and like many Arctic fishes, spawn during the polar winter. Like many polar invertebrates (Clarke, 1980), polar fishes are typically *K*-selected, with large eggs, low fecundities, slow rates of development, and delayed maturity.

8.6 Intersexes

The reader accustomed to mammalian sex life will be much surprised by the extraordinary complexity of the arrangements found in teleost fishes. Many are hermaphrodites of one kind or another, and on coral reefs in particular, it almost seems that hermaphroditism is the rule rather than the exception amongst the acanthopterygians. Indeed, the relatively simple structure of the teleost gonad seems to have readily allowed them to evolve from gonochorism to successive or synchronous types of hermaphroditism. Curiously enough, hermaphroditism is almost unknown in the dominant freshwater group of ostariophysans, although it is common in the dominant marine actinopterygians, especially in families well represented in the coral reef faunas. In the deep sea, on the other hand, intersexuality seems to be largely confined to the Scopeliformes. At mesopelagic levels, synchronous hermaphroditism (when both parts of the ovotestis mature at the same time) seems to be the rule in alepisaurids and notosudids, and the same is true of the benthic chloropthalmids, from 200–6000 m. In the tripod-fishes (*Bathypterois* spp.) of the last family, there are no signs of self-fertilizing hermaphroditism, for Sulak (1977) was unable to find evidence of clone formation in local stocks. Even in species with very wide distributions, there is considerable variety in characters from one region to the next. But the meeting of two ripe individuals may well lead to the fertilization of two batches of eggs. If the energy cost of gonad development in a pair of synchronous hermaphrodites is not much more than in a male and female of a gonochoristic species, in habitats where reproductive contacts are few or production of ova is limited by local constraints, the former reproductive means is clearly advantageous. Both factors may well apply on the deep-sea floor, especially at abyssal levels.

In the mesopelagic zone, individuals of alepisauroid species, which are large-prey eaters, may also be well separated in hunting territories, and most species are without luminous means of signalling to enhance their

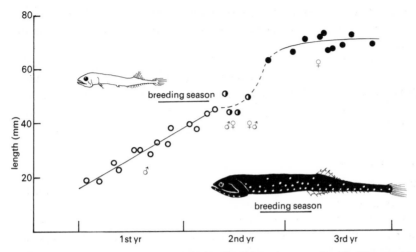

Figure 8.11 Male to female sex change during life of *Gonostoma gracile*. After Kawaguchi and Marumo (1967).

chances of contact. Synchronous hermaphroditism would thus seem to be advantageous to alepisauroids, but it is unknown in the stomiatoid fishes, most of which are mesopelagic. However, the latter have elaborate luminous systems that in many are sexually dimorphic (Marshall, 1979).

Successive hermaphroditism has nevertheless developed in certain bathypelagic stomiatoids. In *Gonostoma gracile*, for instance (Figure 8.11) the males are not only much smaller than the females (males < 60 mm; female 70–90 mm), but change through a hermaphroditic phase into females. There is evidence of similar protandrous hermaphroditism in *Cyclothone microdon* (Badcock and Merrett, 1976). Thus, in both species the males are not discarded after use, but are saved to become females, which is both economical and a means of boosting fecundity in relatively unproductive parts of the ocean.

In shallow seas, most instances of sex-reversing fishes are found in the following perciform families: Pseudogrammidae, Pseudochromidae, Polynemidae (threadfins), Pomacanthidae (angelfishes), Pomacentridae (damselfishes), Scaridae (parrotfishes), Labridae (wrasses), Sparidae (seabreams) and Serranidae (sea-basses). Representatives of all these families live in coral reefs and atolls and the overall distributions of the last three extend into temperate regions.

Certain coral fishes, such as the serranids *Hypoplectus nigricans* (black hamlet) and *Serranus subligarius* are pair-mating synchronous hermaphrodites that develop gonads given mostly to egg production: the testicular part is very small. During mating, hamlets 'trade' eggs, which are released for fertilization in exchange for the opportunity to fertilize those of the other individual. (Fischer, 1981). This behaviour and the great emphasis on egg production means that each of their offspring surviving to maturity will also be mostly female in gonadal emphasis. Thus, their average fecundity will exceed that of hermaphrodites investing equally in male and female components (however, in all known synchronous hermaphrodites the female part of the gonad is much the larger). Moreover, the offspring from a pair mating in a gonochoristic species are likely to be much less numerous than those produced by a pair of synchronous hermaphrodites, with gonads and eggs of the same size. And where population densities are low, the maintenance of a species ought to be best secured through synchronous hermaphroditism. In presenting these and other arguments, Fischer wonders why synchronous hermaphroditism is not commoner among animals.

Successive hermaphroditism seems also to be advantageous in numerous other coral-reef fishes. Wrasses and parrotfishes are dominant groups in reefs and atolls and many are protogynous. Bruce (1979), who studied the parrotfishes of Aldabra Atoll (north of Madagascar), found that one (*Leptoscarus vaigensis*) is gonochoristic: the other 17 species are protogynous hermaphrodites. In the latter forms there may be gonochoristic individuals (without the capacity to change sex), protogynous hermaphrodites (individuals that can change from females to males), primary males (gonochoristic), females (with or without the ability to change sex: at present protogynous hermaphrodites in the female phase cannot be distinguished from gonochoristic females), and secondary males (resulting from sex change in a protogynous hermaphrodite). Both primary and secondary males may be present in a population (diandry) or secondary males only (monandry). There are also colour phases: juvenile (of immature individuals), initial (of small adults), terminal (of larger males) and transitional (between initial and terminal). In sexually dichromatic forms at least some males (in the terminal phase) have a colour pattern differing from that of females in the initial phase.

What are the advantages of sex change? Studies of wrasses seem to fit an extension of Ghiselin's (1969) ideas; that hermaphroditism will evolve in species where one sex benefits from being larger or smaller than the other. Sexual differences in size are related to the size of the breeding group. Thus, in the wrasse *Thalassoma bifasciatum* the expanse of reef occupied limits the size, and presumably genetic fitness, of the breeding groups, but in ways that differ for primary and secondary males. Large secondary males in the small groups of small reefs will have much greater mating success with a limited number of females than will the primary males that lurk in the confusion around a secondary male's territory. On large reefs a few dominant secondary males may be very successful, mating as many as 100 times a day. But most secondary males are no more successful than the smaller primary males that spawn with the large numbers of females attracted to secondary male territories. Thus, secondary males are proportionately commoner and most effective in small breeding groups while the same is true of primary males in larger groups (Sale, 1980).

Lastly, predators are attracted to spawning fishes. Active and brightly coloured primary males are likely to be the most vulnerable. Though the secondary males are also attractively coloured their larger size should help them to evade more enemies. Moreover, secondary males will tend to come from female protogynous hermaphrodites that are most fitted to attain this phase. During their multiple matings secondary males presumably pass on these advantageous characters to their offspring.

CHAPTER NINE

SENSORY SYSTEMS AND COMMUNICATION

Fishes have more senses than ourselves, for they have an elaborate lateralis system to detect near field vibrations, and in addition, a fair number of fishes have modified part of this lateralis system to use it for electro-reception, sometimes using the electroreceptor system for geomagnetic navigation. As compared with terrestrial animals, however, fishes are relatively poorly equipped with proprioceptors. In the cat, for example, sensory fibres account for more than 75% of the axons in the nerves passing to the muscles; most of these go to the muscle spindles. Muscle spindles are absent in fishes, and proprioceptors are known only in a few special cases, such as *Myxine*, teleost barbels, and rays. Remarkably enough, though the ray receptors are simple, they behave rather like the much more complex spindles when their static and dynamic responses are tested physiologically (Ridge, 1977). Probably the absence of an important proprioceptive sense in fishes is related to the relatively insignificant role that gravity plays in their lives, and to the damping of movements by the medium.

This chapter looks briefly at the chief sensory systems in fishes and considers also the use they make of these systems in communicating with each other by the sounds, lights, electric discharges, and pheromones that they produce.

9.1 The acoustico-lateralis system and sound production

The acoustico-lateralis system enables fishes to respond to vibrations in the water, to gravity, and to angular accelerations; originally it arose as a system for detecting low-frequency vibrations at short range, and its other functions appeared later in chordate phylogeny. Amphioxus has apparently no analogue of the acoustico-lateralis system, although one is present in certain tunicates.

Fishes detect vibrating objects (such as other swimming fishes) with the hair cell receptors of the ear and lateral line. Simple vibration receptors, like those of chaetognaths or sea-squirts, consist of long stiff non-motile cilia surrounded by a corolla of microvilli, but the hair cells are interestingly different in design (Figure 9.1). There is a kinocilium, which has the usual $9 + 2$ internal tubular array, but it lies to one side of a stack of microvilli (misleadingly called stereocilia for they do not have internal tubules) so that the hair cell is morphologically asymmetrical.

At their bases, the hair cells receive chemical synapses (perhaps utilizing amino acids as the transmitter) from central sensory cells, but in addition to this afferent innervation, they also receive *efferent* synapses, whose action 'switches off' the hair cells (Figure 9.1). The physiological role of such efferent synapses (not found at the bases of electroreceptors) is unknown, but may be linked to damping hair cell responses to the movements of the fish itself as it swims.

The kinocilium and stereocilia (the term is too embedded in the literature for us to abandon it!) project into a gelatinous cupula, and the hair cell responds to deformations of the cupula along the axis of symmetry

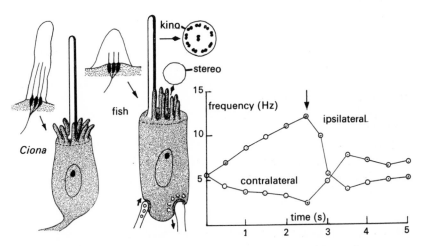

Figure 9.1 Vibration receptors. Left: comparison of ascidian and fish cupular receptors. Transverse sections of the kinocilium (kino) and a stereocilium (stereo) are shown above the fish hair cell. Note afferent and efferent synapses at the base. Right: responses of horizontal semi-circular canal maculae in *Raja* showing changes in impulse frequency during rotation (stopping at arrow). After Lowenstein and Sand (1940).

of the cell, i.e. to movements which bend the kinocilium and stereocilia towards and away from the position of the kinocilium. By analogy with other ciliated sensory cells, we might have expected that the cell responded to bending of the kinocilium, rather than to movements of the stereocilia, but at least in amphibians, this is not the case. Elegant experiments by Hudspeth and Jacobs (1979) on the large hair cells of bullfrogs have shown that hair cells deprived of the kinocilium still respond to displacements of the stereocilia in the same way as intact haircells, as do occasional haircells in which the kinocilium is naturally lacking. This intriguing discovery suggests that there must have been a change in the way in which hair cells are stimulated from an original situation where movement of the cilium was the effective stimulus, to that where the cilium is redundant, and the effective stimulus is movement of the stereocilia. It would be interesting to examine the large hair cells of holocephali to see whether this change has taken place in fishes. In any case, the kinocilium provides a convenient marker of the polarity of the hair cell for the scanning microscopist.

In the ray labyrinth, Lowenstein and Wersall (1959) found that the hair cells of the horizontal canal had their kinocilia oriented towards the utriculus, whereas in the anterior vertical canal, the hair cells had their

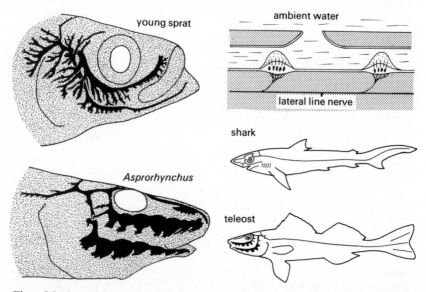

Figure 9.2 Lateral line system. Left: lateral line canals of young sprat and *Asprorhynchus*. Right: arrangement of maculae and cupulae in lateral line of shark. Below: lateral line canals of shark and teleost. After Jakubowski (1974) and Blaxter *et al.* (1981).

kinocilia oriented away from the utriculus. By recording from these two hair cell groups whilst rotating the labyrinth appropriately, they were able to demonstrate that these hair cells were inhibited when the stereocilia were moved towards the kinocilium, and excited when they were moved away (Figure 9.1). Although the specially simple hair cell arrangement in the labyrinth is not present elsewhere, so that similar tests have not been possible, it seems very probable that all hair cells operate in the same way; whether their resting discharge is augmented or inhibited depends on the direction the stereocilia are bent. By arranging the polarity of the haircells suitably, therefore, a fish can detect directional movements of the cupula into which they protrude, and can distinguish between movements in either direction along the same axis. It seems clear that the lateral line cupular organs first arose as free external receptors, a suitable starting point being the cupular vibration detectors of tunicates where the receptors are not morphologically polarized (Bone and Ryan, 1978), and later became enclosed in lateral line canals.

Open lateral line canals are found in holocephali but usually they are closed and only open to the surrounding water by small pores at intervals (Figure 9.2). Presumably this arrangement shields the cupulae (and receptors) from unwanted deformations (e.g. 'noise' produced by currents and by the movements of the fish itself). External neuromasts and cupulae are found in many fish larvae and in relatively immobile fishes living in 'quiet' environments such as caves and the deep sea.

To understand how vibrations from objects in the water affect the hair cells of the lateral line and ear, we need first to glance briefly at the properties of a vibrating source in a fluid. Such a source produces two different effects simultaneously. As it moves, it displaces the adjacent water, and this direct or near field effect is not propagated, declining as the cube of the distance from the source. At a given distance from the source, the magnitude of the displacements produced is inversely dependent on frequency. At the same time, movements of the source alternately compress and rarefy the adjacent water, thus giving rise to propagated pressure or sound waves, which decline as the square of the distance from the source. The pressures produced are directly proportional to frequency (at a given amplitude of vibration), and we should therefore expect that this far field effect would be most important at greater distances, and at higher frequencies than the near field effect. The near field effect produces greater displacements than the far field up to 0.2 of a wavelength from the source: at all greater distances the far field effect produces larger displacements.

With this in mind, we should guess that at low frequencies where the wavelength is long (at 100 Hz it is 15 m) the near field effect might be most important to the fish, and most easily detected at useful ranges, whilst at high frequencies, the fish would more easily detect the far field effect, particularly if it has devised an effective pressure-movement transducer by coupling the hair cells mechanically to gas-filled spaces that are more compressible than water. A variety of teleosts have succeeded in doing this by arranging connections of one kind or another between the swimbladder and the ear, and fish of this kind (like the Ostariophysi or herrings) are much more sensitive to high frequency than are fishes without such devices.

9.1.1 The lateral line

The lateral line receptors in most fishes lie in canals on the head, and in a canal extending along the body (Figure 9.2), arranged in small cushions spaced along the lateral line canals. The arrangement of the lateral line of the body strongly suggests that the receptors are laid out to provide a long baseline, and in several quite different sorts of fish (for example, holo-cephali and macrourids), there is an elongate thread-like tip to the tail which much extends the length of this baseline. A similar elongation is found in many electric fishes presumably for a similar reason.

Now, by comparing the responses of different hair cells along such a baseline, the fish should be able to use the receptors to locate the source of vibrations. Lateral line neuromasts respond up to frequencies around 200 Hz (where the wavelength is 7.5 m), and appear to be used by the fish to detect the near field effect. The sensitivities of the receptors are remarkable, threshold displacement being only 1–2 nm (Harris and van Bergeijk, 1962; Banner, 1967). Calculations by Pumphrey (1950) suggested that the elongate midwater trichiurid *Aphanopus* (Figure 3.12) could locate the fishes it preys on at distances up to 32 m, using the lateral line to detect the near field effect produced by movements of the prey. At such sensitivity, the near field effects of the prey would be lost in noise produced by any near field effects caused by body oscillations of the predator. *Aphanopus* avoids this problem in an interesting way. It swims slowly by sculling itself along with the minute forked tail, keeping the body rigid (there is a flexible joint in the vertebral column at the base of the caudal peduncle). The prey can thus be stalked at low speed, using the lateralis system, and only when in range of the large sensitive eyes does the fish extend its dorsal fin and oscillate the body in an anguilliform way. Other trichiurids (such as *Trichiurus*) have no caudal fin, and the long dorsal fin is enlarged. The lateral line in these members of the family passes along the body in an unusual ventral position, and it seems that such fishes also use the system to stalk their prey, but move slowly by undulating the dorsal fin; the ventral position of the lateral line removing it as far as possible from disturbances caused by the dorsal fin. This second method of propelling an elongate body acting as a support for the lateralis baseline parallels that found in the freshwater electric knifefishes, where the detection system is different, but its requirements the same.

Trichiurids are obviously highly specialized for lateral line detection of near field effects—they have enormous acoustico-lateralis lobes in the brain (Muir Evans, 1940); the evidence at present suggests that more normal fishes are able to detect near field effects at ranges up to a metre or so, and use their lateral lines to locate sources vibrating at frequencies up to 200 Hz, such as the beating tails of other swimming fish.

9.1.2 Sound reception

At higher frequencies, where the far field or pressure effect is dominant, the hair cells of the ear are used. Fishes without swimbladder-ear connections such as tuna (Iversen, 1967) or cod (Buerkle, 1968) have an upper limit of hearing at around 400 Hz, when tested by cardiac conditioning. Similar tests on fishes which have pressure-displacement transducers gave upper limits up to 7 kHz (von Frisch, 1938), and maximum sensitivities around 1.5 kHz (Enger, 1967). How are these pressure-displacement transducers designed?

In the ostariophysi, a chain of small ossicles of vertebral origin link the swimbladder with the sacculus at the base of the ear (Figure 9.3). These ossicles were described and their function understood by Weber as long ago as 1820, and they are rightly still termed the Weberian apparatus, though some of the names he gave to the separate ossicles have now been changed. Figure 9.3 shows how the system works. Changes in swimbladder volume, as sound waves pass through the fish, rock the tripus, and this movement is transferred via the intercalarium and scaphium to the claustrum which abuts on to the perilymphatic sinus impar. This is directly linked to the wall of the endolymphatic transverse duct joining the sacculi of either side, and so as the swimbladder moves, so too does endolymph in the sacculus, moving the sagittal otolith and stimulating the hair cells of the saccular macula. A surprising number of non-ostariophysan fishes have simpler connections of the swimbladder or some air-filled cavity with the endolymph of the inner ear (via a thin deformable part of the saccular

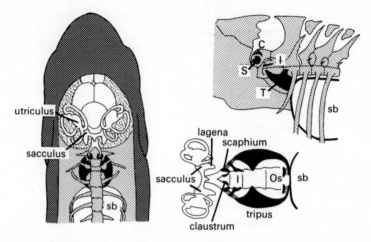

Figure 9.3 Ostariophysan ear ossicles. Left: dorsal view of goldfish; upper right: lateral view of young goldfish; lower right, dorsal view of adult goldfish ossicle series. sb, swimbladder; bas.occ., basoccipital; s, scaphium; i, intercalarium; t, tripus; Os, suspensory bone. After Chraniloff (1929), Watson (1939), and von Frisch (1938).

or lagenar wall). In mormyrids, anterior branches from the swimbladder become cut off in ontogeny, and form isolated sacs (each with its own rete mirabile to fill it (see Chapter 4) lying amongst the semicircular canals (Figure 9.4). Anabantid fishes have an accessory respiratory chamber above the gills (Chapter 5) connected to the inner ear by a thin-walled window. They are normally sensitive to high frequency sound, but if the respiratory chamber is filled with water (instead of air as it usually is) the fish loses this sensitivity.

The most carefully analysed system is that of clupeids, described in the herring and sprat by Denton and Gray and colleagues (1977, 1981). The anatomy of this system is shown in Figure 9.4 and Figure 9.5. A thin deformable membrane separates the perilymph of the utriculus from a special gas-filled auditory bulla. This is kept full of gas by thin tubes leading from the swimbladder. Pressure vibrations of the gas in the bulla move the membrane and thus vibrate the utricular maculae against the utricular otolith (Figure 9.4). By introducing electrodes into the utricular endolymph of a fish set up in a pressure chamber, Gray and Denton were able to record the electrical responses of the hair cells in the utricular maculae to pressure changes.

In response to a sinusoidal pressure wave, microphonics of double frequency were recorded, some cells evidently being stimulated as the

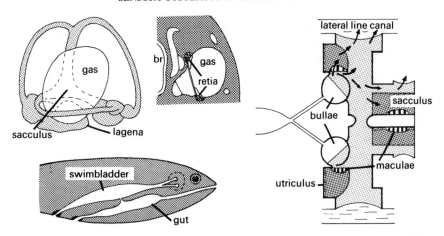

Figure 9.4   Gas-filled bullae and inner ears in mormyrids and clupeids. Upper left: mormyrid ear showing gas bulla applied to outside of sacculus; in transverse section of the head, the retia filling the bulla can be seen. Lower left: connections of the herring or sprat swimbladder. Right: schematic diagram showing relations of gas-filled bullae with inner ear. Fine stipple: perilymph; coarse stipple: endolymph. After Stipetic (1939) and Denton and Gray (1979).

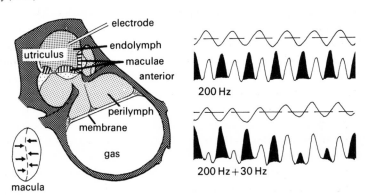

Figure 9.5   Responses of sprat ear. Left: sagittal section (posterior to left) showing gas-filled region of bulla separated from perilymph (fine stipple) by membrane. The recording electrode lies in the utricular endolymph (coarse stipple) above the maculae (striped). Dispositions of the two orientations of kinocilia in the middle utricular macula are seen to left. Right: responses of the utricular maculae hair cells to sounds (sinusoidal pressure waves). After Denton and Gray (1979).

pressure increased, others as it decreased (Figure 9.5). In view of what we know about the asymmetry of hair cells it is reasonable to equate these two types of response with the two orientations of hair cells in the utricular middle macula which lie facing the macular midline. Denton and Gray found rather flat responses over a wide frequency range (40–700 Hz), but when the pressure stimulus was a mixture of two sine waves of different frequency, the result was very striking (Figure 9.5). The responses of the receptors responding to compressions, and those to decompressions depended on the pressure changes above and below mean pressure, and so the phase relations and the two frequencies are much more obvious in the electrical response than in the pressures which gave rise to them. More complex pressure stimuli where two frequencies are mixed with differing phase lags are very easily distinguished by the utricular responses.

The striking thing about these observations is that they give strong support to the idea that herring are interested in *different* features of sounds than we are. In our cochlea, there is no frequency doubling, and to detect the frequency of a sound we need to hear several cycles; shorter sounds of different frequencies sound like clicks, and we cannot estimate their pitch. The herring is able to determine frequency using a single half cycle alone, and (like other fishes with specialized pressure-displacement transducers) can respond very rapidly to appropriate frequencies using the Mauthner pathway (Chapter 10).

The herring ear, like that of other fishes, has dense calcium-loaded otoliths lying over the maculae, and it is the movement of the hair cells against these otoliths which stimulates them during vibration perception. The density of the otoliths makes them operate not only during vibration detection, but also in gravity reception (as the fish tilts from the horizontal, the otoliths slide downhill and so stimulate their maculae), and they also act as accelerometers. The otoliths in bony fishes are often curiously shaped, fitting into the spaces above the maculae very exactly, but in agnatha and elasmobranchs the otoliths are amorphous and a dense jelly is loaded with small particles of calcium salts, or even sometimes (in the bottom-living elasmobranch *Rhina*) with sand grains! Fänge (1982) considers how sand may enter the inner ear.

9.1.3 Sound production

Fishes produce sounds in various ways, by grinding their teeth, by rasping spines and fin rays, or by burping, farting or gulping air. Some of these mechanisms may involve the swimbladder as a resonator—swimbladder deflation changes the sounds produced by grunts (Pomadasyidae) when they grind their pharyngeal teeth—but the swimbladder is directly involved in many fishes where special muscles cause it to vibrate and thus operate as an underwater speaker. These muscles may be intrinsic to the swimbladder wall, or insert at one end on the swimbladder, at the other on the skeleton. Sounds produced by vibrating the swimbladder may be at high frequency (400 Hz) and the swimbladder drumming muscles are correspondingly specialized for rapid contractions with large SR volume and small diameter fibres.

Sounds produced in this way may be sufficiently intense to be heard above water, and since drumming muscles are only present in the males in some species, there is good reason to suppose that the sounds produced are important during courtship. Curiously enough, although many catfishes generate sounds, these are usually well below the maximum frequency to which the ostariophysan ear responds, and a similar puzzle is provided by herring, which also respond to high frequencies although most of the sounds in their environment are 100 Hz or less. It hardly seems probable that herrings have developed the complex high frequency sound detection capability simply to be able to listen to the high frequency sounds emitted by marine mammals hunting for them; perhaps they use this capability to detect sounds made by crustacean prey. Another possibility is that they eject small bubbles of gas via the anal opening of the swimbladder (Figure 9.4) thus producing high-frequency sounds for communication.

Sciaenids drum as they feed (which they do diurnally) and possibly here the drumming sound produced by one fish as it feeds attracts others of the species to a source of food, but as Marshall (1962) points out, this may alert predators, or scare prey as well as signalling to others of the same species. The growling and hooting sounds of the batrachoid *Opsanus tau* seem to be used as warnings and as indications of territories. Little is definitely known, however, of the functions of sound production in many fishes.

9.2 Electroreceptors and electric organs

A surprising number of fishes in different groups possess electroreceptors of various types, and many of these also possess electric organs. The sensitivity of some of these receptors is indeed extraordinary—they are able to detect electrical fields of 0.01 µV/cm, or 1 mV/km! The reader who has struggled to record signals 100 or 1000 times larger with high-quality electrophysiological equipment will appreciate this performance. Not only do such receptors enable their owners to detect the small steady d.c. fields of living prey buried in the substrate (Figure 9.6), but sensitivity of this order is quite sufficient to permit the fish to detect the fields arising as it

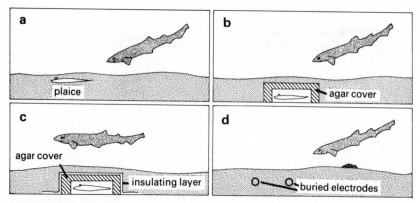

Figure 9.6 Electroreception in dogfish. (*a*) The dogfish attacks a buried plaice (*Pleuronectes*), detecting the d.c. field it generates. (*b*) If the plaice is within an agar chamber 'transparent' to current, it is also attacked. (*c*) If the agar chamber is covered with an insulating plastic layer, it is not detected. (*d*) Buried electrodes producing a field like that of the plaice are attacked. After Kalmijn (1971).

swims through the magnetic field of the earth, and so to orient itself geomagnetically.

Electroreceptors of this kind, which respond to d.c. fields by changes in their tonic discharges, are found in all elasmobranchiomorph fishes, in a variety of teleosts, in sturgeons and *Polyodon*, and in Dipnoi. It seems probable that they also occur in *Latimeria*, but have not yet been described (our preliminary hunt around the snout has so far been unsuccessful!). Some of these fishes also possess electric organs, as do rays, for example, but the majority of them do not. Electroreceptors of a different kind, which respond to high frequencies, and are insensitive to d.c. fields, are found in several families of specialized freshwater fish, all of which possess electric organs. This second type of receptor produces phasic discharges in response to high-frequency signals and is used in conjunction with the electric organ discharges to detect prey, for social signalling, and for electrolocation. Such fishes also have tonic receptors.

How are these different kinds of electroreceptors designed, and how do they operate?

9.2.1 Tonic ampullary receptors

This type of receptor consists of a canal leading from the surface of the skin to an ampulla in whose walls a group of sensory cells are embedded (Figure 9.7). Only the tip of the sensory cell is exposed to the lumen of the

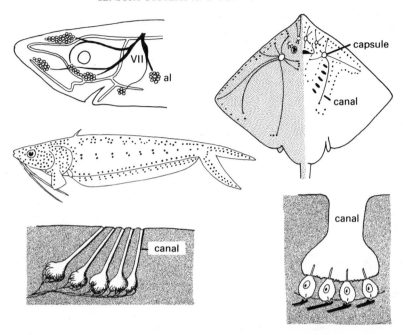

Figure 9.7 Ampullary electroreceptors. Upper left: groups of ampullae (al) on head of shark. Mid-left: distribution of ampullary organs on the catfish *Kryptopterus*. Bottom left: groups of ampullary organs from shark snout. Upper right: upper and underside of ray showing openings of ampullary organs, and capsules into which canals collect. Only a few canals are indicated. Bottom right: Ampullary organ from sturgeon snout, the sensory cells are ciliated. After Murray (1967), Jorgensen (1980), Bennett (1971).

ampulla; at its base there are synapses with fibres from lateral line nerves. Both tonic and phasic receptors are modified lateralis receptors, but interestingly, unlike the lateralis receptors, no efferent innervation of the receptors has been reported.

Ampullary receptors are widely distributed on the body in electro-sensitive teleosts around the head in sharks and over the upper and lower surface of the 'wings' in rays (Figure 9.7). In sharks and rays the canals lead to ampullae which are grouped within capsules, but in teleosts, they are separate from each other. The canals are filled with low-resistance jelly (similar in ionic composition and resistance to sea water), and their walls are of high resistance (30–100 times higher than that of the nerve myelin sheath), so that they are effectively ideal submarine cables with zero leakage conductance and inductance, ending in an open circuit at the ampullary end of the canal. Thus the ampullary sensory cells are exposed at their apices to the same potential as that in the sea at the opening of the canal.

At their bases (which lie outside the ampulla, and in sharks and rays, within a thick-walled capsule), the sensory cells will be at the potential of the interior of the fish at that point. Presumably, the tonic resting discharge of the receptor cell results from a steady 'trickle' of neurotransmitter at the basal synapse, and changes in potential between the apex and base of the cell alter the rate of transmitter discharge.

Ampullary organs in freshwater and marine fishes differ in the lengths of their canals (Kalmijn, 1978). Conditions are very different in the two media: in the sea, the body fluids have a higher resistivity than sea water for they are more dilute (Chapter 6), and the skin is only a moderate insulator. Figure 9.8 shows the result of this in a uniform field; the inner part of the sensory cell lies surrounded by body fluid nearly at the same potential as that of the sea water just outside the body at that point. The apex of the cell is at the same potential as that at the opening of the canal, hence the longer

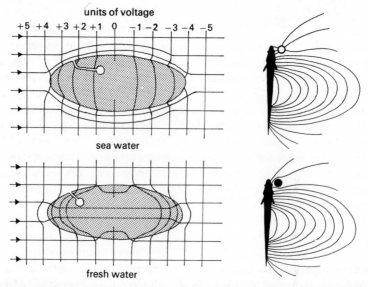

Figure 9.8 Passive and active electroreception. Left: fields around marine and freshwater fish with ampullary organs, anode to left. The freshwater fish has high skin resistivity, and needs only short ampullary canals, whilst the marine fish has lower skin resistivity and the canals are longer. Right: Current lines of the fields generated on one side of a freshwater electro-locating fish. In the upper diagram the field is distorted by an object of high conductivity (such as a living prey); below, by an object of low conductivity (such as a rock). After Bullock (1973) and Lissmann and Machin (1958).

the canal, the greater difference in potential there will be (in a uniform field) between the apex and base of the sensory cell.

It is interesting that the ampullae in elasmobranchiomorphs are collected to lie in capsules, for this means that the cell bodies of sensory cells whose apices are connected to canals opening far from each other on the body surface, are as near isopotential as possible, and thus the fish will be able to compare most accurately the differences in potential along its body.

In fresh water, the body fluids have a lower resistivity than the medium, and skin resistance is high, so that in a uniform field, the body fluids are almost at isopotential throughout, and the bases of the receptor cells are at the same potential as the external field about the middle of the fish (Figure 9.8). Here there is no need for long canals leading to the ampullae, and as Szabo *et al.* (1972) found, the freshwater stingray *Potamotrygon* of the Amazon has very short canals, as have other freshwater electric fishes with ampullary organs.

9.2.2 Phasic tuberous receptors

A second class of phasic electroreceptors are differently arranged to the tonic receptors. There is no canal, but instead a plug or tube of cells lies above the receptor cavity, presumably forming a low resistance pathway from the surface to the electroreceptor cavity. Within the cavity, the receptor cells protrude, attached only at their bases, their surface being much increased by closely packed microvilli (Figure 9.9). This huge surface acts as a series capacity, whilst the inner basal membrane of the sensory cell is electrically excitable, and may even generate spikes. In consequence of the series capacity, receptors of this sort adapt rapidly, and are suited for reception of the high frequency discharges from the electric organs of the fish.

9.2.3 Electric organs

With the exception of the freshwater sternarchids, where the electric organs are derived from modified nerve fibres, all electric organs are modified from striated muscle fibres, and consist of stacks of flattened cells, innervated on one side (Figure 9.9). This arrangement sums the small potentials arising from membrane depolarizations (as in normal nerve and muscle cells), thus giving rise to much larger external potentials. In a few fishes, such as the marine *Torpedo*, and the freshwater eel *Electrophorus*

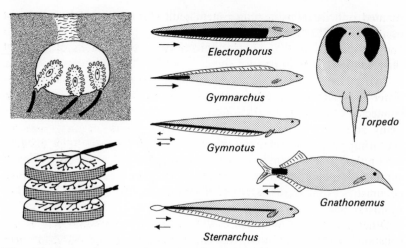

Figure 9.9   Active electroreception and electric organs. Upper left: tuberous mormyrid receptor (note large surface area of receptor cells within cavity). Lower left: stacked modified muscle fibres and their innervation in the electric organ of *Torpedo*. Right: positions of electric organs in different fishes—the arrows indicate directions of current flow. After Bennett (1971), and Szamier and Wachtel (1970).

and catfish *Malapterurus*, the electric organs generate large potentials, sufficient to stun their prey. *Electrophorus* produces 500 V, *Torpedo* around 50 V (measured in air), and such discharges may also be used as a very effective defence. Most electric fishes however, produce much smaller discharges than this, and the function of such weak electric organs (generating a volt or less recorded in air) remained a puzzle, until Lissmann (1958) suggested in a classic paper that they functioned in electrolocation. Lissmann's view was that the fish could detect (with its electroreceptors), distortions of the field around it generated by its electric organs, and hence infer the size, distance, shape, and conductivity of objects producing such distortions in the field. He pointed out that freshwater species of weakly electric fishes lived in very turbid waters, where vision would be of little value, but where electrolocation would be useful.

Subsequent experiments by Lissmann and Machin and others have confirmed that weakly electric fishes indeed use electrolocation to examine their surroundings, and can distinguish objects of different impedance from the medium by the distortions they produce in the field generated by the electric organs (Figure 9.8). Analysis of the electroreceptor signals is obviously simplified if the fish keeps its body rigid, and as Lissmann pointed out, fishes using this system scull slowly around by means of oscillations of a dorsal or ventral unpaired fin, so the body remains rigid and the spatial orientation of the receptors *vis-à-vis* the electric organ and the field it produces stay the same.

Two kinds of signals are produced by different weakly electric freshwater fishes, and can be simply detected by electrodes in the water, connected to amplifiers and speakers, as in the displays of *Electrophorus* in many zoo aquaria. As well as the powerful electric organ it uses for capturing prey, *Electrophorus* has weak electric organs which produce brief pulses at low frequencies, as do mormyrids and some gymnotids. If stimulated in any way, these increase in frequency, and under some circumstances may cease altogether, presumably to camouflage the fish against others which could detect it electrically.

Other gymnotids and *Gymnarchus* emit pulses continuously at a very constant high frequency, so that their signals are a continuous near sinusoidal wave rather like 50 or 60 cycle a.c. but up to the extraordinary frequency of 2000 Hz!

The functional significance of the difference between the pulse species and the wave species is not yet clear, but evidently both kinds of signals can be used not only to examine the environment, but also to communicate with other fish possessing electroreceptors. In the tropical rivers and lakes where these fishes are found, many species occur together, as well as other individuals of the same species, so although the fishes could identify each other by their characteristic discharges, there are the same kind of interference problems as occur when echo-locating bats are flying around together. Szabo *et al.* (1972) found that species producing a steady wave discharge change frequency whenever another fish (or an electrical signal source) with a frequency within 1–2 % comes within range.

This jamming avoidance response always operates in the correct direction (to increase the difference between the frequency of the intruder and the fish's own frequency), and shows a strikingly precise control of the firing frequency of the electric organ. Some fishes of this kind have been found to show standard deviations of only 0.1 µs over many thousands of cycles for the period between successive discharges. This astounding precision means that a number of such fishes can operate successfully in close proximity without interfering with each other.

The weakly electric fishes considered above all live in fresh water. Marine rays are also weakly electric, with caudal electric organs. The function of their electric organs remains mysterious, for they only emit pulses very rarely and evidently do not use the system for electrolocation.

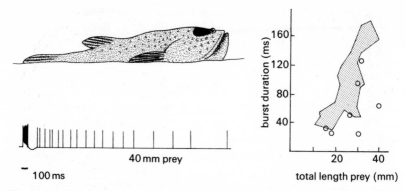

Figure 9.10 Electric discharges in the only marine electric teleost. Left: stargazer (*Astroscopus*) half-buried in sand; below, discharge after catching prey. Right: relationship between discharge and prey length. Open circles, misses. After Pickens and McFarland (1964).

Perhaps the electric organ discharges of rays (the pulses differ in shape and duration in different species) may be employed for recognizing other members of the same species. The bizarre electric organ of another marine group, the stargazers, which is modified from their eye muscles, presents another very curious puzzle—*Astroscopus* (Figure 9.10) lies on the bottom awaiting small fishes which are engulfed as the stargazer opens its mouth and vacuums them in. During this process (which takes only 150–300 ms), the electric organs fire a burst of pulses at high frequency, followed by a train of discrete pulses lasting a second or so (Pickens and McFarland, 1964). The discharges are less than a volt (measured at the mouth) and insufficient to stun the prey. Remarkably, the duration of the burst which occurs as the mouth opens is correlated with the length of the prey fish (Figure 9.10). It is almost as if the discharge signals to other stargazers the pride that the captor feels in the size of its meal! Whatever its real significance, the stargazer electric discharge is certainly not used for electrolocation, and Picken and McFarland's work has provided us with an amusing puzzle.

9.3 Vision and photophores

Very few fishes live in total darkness. In the clearest oceanic waters, sufficient light for vision with sensitive eyes reaches down to around 1100 m (the limit for bright moonlight is around 600 m), but below these depths most fishes have large and complex eyes, for the natural darkness is

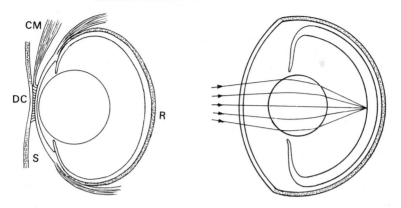

Figure 9.11 Left: eye of lamprey showing corneal muscle (CM), retina (R) and sclera (S). Right: ray paths through fish lens. After Duke-Elder (1958).

relieved by the flashes and glows of bioluminescence. Only in caves is the visual system useless, and there some 40 spp. of blind fishes are known, belonging to several families. Fish eyes are much more varied in design than those of terrestrial animals, partly because the light regimes in which fishes live are more varied, and partly because the lateral position of the eyes on a head placed on an inflexible neck requires special modifications for a large field of binocular vision. In some cases too, there are air–water interface problems to be overcome, as in flying fishes or in freshwater fishes preying on insects above the water, like the archer-fish *Toxotes*.

In cyclostomes, although the eyes of hagfishes are somewhat reduced and lie buried below the skin, those of adult lampreys (Figure 9.11) are like those of gnathostomes, and in essentials, all fish eyes are of this general arrangement.

9.3.1 Optics

The refractive index of the cornea and ocular fluids is similar to that of the water, so that refraction takes place almost entirely in the lens, which is usually nearly spherical and has a short focal length (in teleosts very close to 2.55 × the lens radius). The overall refractive index is remarkably high (around 1.69), but the quality of the lens is very striking. A fresh teleost lens, although nearly spherical, yields an image almost free of aberration—a glass marble under similar circumstances provides an appallingly aberrated image. How is the fish lens so well corrected? The physicist James Clerk Maxwell was much impressed by the perfect imaging properties of the fish lens, and deduced in 1853 the general properties of such lenses, but the detailed optics were only worked out a century later by Fletcher, Murphy and Young (1954). The refractive index of the lens varies across its diameter in such a way that rays passing through the lens follow curved paths (Figure 9.11), and form a sharp image, free from spherical and chromatic aberration.

There is still argument about the way in which fishes accommodate to examine close and distant objects, and whether at rest the eye is focused close to or far from the cornea. We accommodate by changing the focal length of the lens (altering its radius of curvature by permitting the lens to bulge) but almost all fishes instead change the distance between the lens and retina by moving the lens backwards and forwards along the optical axis. Even in fishes belonging to the same group, the mechanism of accommodation can be different, and is a most fascinating study.

Lampreys accommodate by an extraordinary and unique mechanism, for the cornea is attached to a circular muscle (of myotomal derivation) (Figure 9.11) which contracts to flatten the cornea and so pushes the lens towards the retina to focus distant objects. This unusual system seems to be assisted by changes in lens–retinal distance brought about by contraction of the extra-ocular muscles, and even by lateral movements of the head, but these additional mechanisms of accommodation are not yet properly understood.

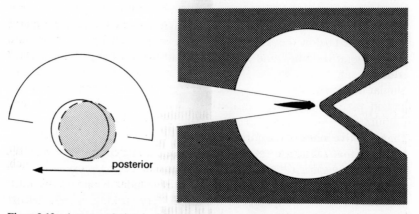

Figure 9.12 Accommodation in fishes. Left: trout eye showing lens movement from rest to fully accommodated (stippled). Right: horizontal visual field (eyes unaccommodated) in focus stippled. White areas to either side of the fish are within the locus of the near point, the anterior white area beyond the far point, and the white area behind the fish is the shadow of the body. After Pumphrey (1961) and Somiya and Tamura (1973).

In elasmobranchs, the classical view was that the lens is moved towards the cornea by contraction of a protractor lentis muscle, but histological examination of this supposed muscle has not demonstrated muscle fibres, and electrical stimulation experiments have yielded negative results, so that it is unclear whether elasmobranchs can accommodate. In some stingrays, the distance between the retina and the lens (which is not quite spherical, and has a lower refractive index than that of teleosts) varies around the eye, so possibly different regions of the retina are used to focus near and far objects so that accommodation is a static rather than a dynamic process. No one who has ever been in the water whilst being circled by an inquisitive pelagic shark such as *Lamna* can doubt that some elasmobranch eyes are used during the examination of prey, and that accommodation is probable!

The teleost lens can be moved relative to the retina by the retractor lentis muscle, and when this muscle is stimulated electrically, the lens is moved in most species towards the tail of the fish, in some obliquely downwards, and in yet others, obliquely upwards. This seems a rather peculiar system, for we might expect that the retractor lentis muscle would operate to move the lens towards the mid-line of the fish, into the eyeball. In a very few fishes (for example goldfish) it seems to do just this, but in the majority it does not, and Pumphrey (1961) suggested that the situation is a consequence of the shape of the eyeball; the retina is not concentric with the lens, but is somewhat ellipsoidal (Figure 9.12). Considering those fishes where the lens muscle pulls it rearwards (as in the trout), Pumphrey suggested that the eye at rest was shortsighted anteriorly, and far-sighted laterally, so that the field in focus would look in horizontal section as in Figure 9.12. Backward movement of the lens on accommodation would thus little alter the field laterally, but would enable the fish to focus distant objects anteriorly. Similar arguments apply where the direction of lens movement is oblique, and in marine fish with this method of accommodation, Somiya and Tamura (1973) have calculated that objects from infinity down to within about half the standard length can be focused.

Freshwater fish can accommodate less, as we might expect, since in the more turbid waters of rivers and lakes, the farthest limit of useful vision is likely to be only around 1–2 m, whereas the oceanic waters, this limit is around 30 m.

9.3.2 Tubular eyes

Normal teleost eyes of this kind have been strikingly modified in many mesopelagic fish. Eleven families (including hatchetfish and giganturids)

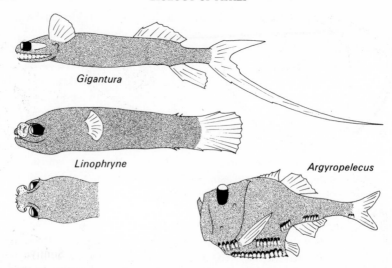

Figure 9.13 Mesopelagic fishes with tubular eyes. Note that *Linophryne* looks through its transparent olfactory organ. After Brauer (1908).

have independently redesigned their eyes to enlarge the field of binocular vision, and have tubular eyes which point forwards or upwards, with the optical axes of both eyes more or less in the same direction (Figures 9.13, 9.14). Such tubular eyes were at first supposed to act as telescopes, but the distance between the lens and main retina is the same as in the normal teleost eye (Figure 9.14), and the tubular eye functions (along the visual axis) just as in normal teleosts. The polar diagram of natural light in the sea (see Figure 9.19) is such that animals have to attempt to solve the difficult camouflage problem of avoiding being silhouetted against down-welling light when viewed from below. It is not surprising that most mesopelagic fishes with tubular eyes have them arranged to look upwards. Those like giganturids where the tubular eyes look forwards (Figure 9.13) probably adopt an oblique attitude in the water when at rest (owing to the relative positions of the centres of gravity and buoyancy, see Chapter 3), and so actually are usually also positioned to look upwards.

Tubular eyes enable their possessors to achieve good binocular vision in one direction, but the retina outside the main portion is too close to allow other than unfocused movement detection. Some fishes with tubular eyes have surmounted this in a most curious way, by developing accessory retinae or even accessory refracting devices so to obtain images from light entering the eye outside the main axis of the eye (Figure 9.14). The

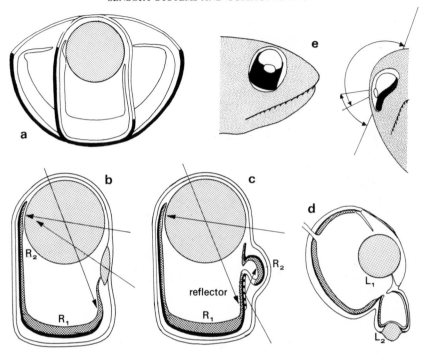

Figure 9.14   Tubular eyes and their modifications for lateral vision. (*a*) Tubular eye of the sternoptychid *Argyropelecus* superimposed upon normal fish eye; (*b*) *Scopelarchus* showing accessory scleroid lens; (*c*) the opisthoproctid *Dolichopteryx* with reflector (R) and accessory globe; (*d*) the opisthoproctid *Bathylychnops* with scleroid lens and secondary globe; (*e*) anterior view of *Scopelarchus* showing its fields of view—lower field of view is via scleral lens and accessory retina, upper main retina (note wide dorsal binocular overlap) L_1, L_2: main and accessory lenses, R_1, R_2: main and accessory retinae. After Munk (1966) and Locket (1977).

argentinoid *Dolichopteryx* has an accessory retina illuminated by light which has passed through the lens and is then reflected from the argentea into the accessory retina. Scopelarchids and evermannellids have no accessory retina, but lateral light falling on the upwardly directed tubular eyes is focused on to the main retina by accessory refracting lens pads and ocular folds (the former intra-ocular, the latter formed by a modification of the adipose eyelid). Evidently, inefficient lateral vision is a high price to pay for efficient binocular vision, and is the reason for these extraordinary modifications of the original tubular eyes. Owls, which also have tubular eyes, which cannot be moved in their sockets even with pliers (Walls, 1942) compensate for poor lateral vision by having extremely mobile heads.

9.3.3 Aerial vision

Quite different modifications have arisen in teleosts which are adapted for vision in air, such as mudskippers, or the remarkable 'four-eyed' *Anableps* which has divided the retina into two parts, and by using an aspherical lens, can focus both in air and in water. In air, the curved cornea is refractive, and flying fish and some shore-living clinids ingeniously avoid being short-sighted in air by arranging two flat 'windows' in the cornea to enable them to look downwards at the water surface.

9.3.4 Reflecting tapeta

In most elasmobranchs, in holocephali, in sturgeons and *Latimeria*, as well as in a few teleosts, light passing through the visual cells is reflected back again by a tapetum backing the eye (Figure 9.15). The eyes of these fishes shine, as do those of cats; those of deep-sea sharks hauled aboard ship shine a superb greenish-blue colour. Such choriodal tapeta consist of layers of reflecting cells packed with thin platelets of guanine, like those found in the teleost scale. The platelets are arranged at the appropriate angles to reflect light back into the retina along the long axis of the rods

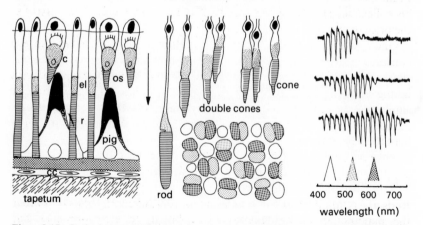

Figure 9.15 Retinal organization and function in actinopterygians. Left: sturgeon retina showing visual cells, pigment cells, and tapetum. cc; chorio capillaris; pig, pigment cell; el, ellipsoid; r, rod; c, cone; os, outer segment. Middle: cell types in the teleost retina. Below: cone mosaic (white, blue; fine stipple, green; coarse stipple, red). Right: response spectra of single carp cone types when illuminated by 0.3 s light flashes 20 nm apart across the spectrum. Scale bar: 2 mV. The three peaks below are absorption maxima found by microspectrophotometry. After Walls (1942), Engstrom (1963), Tomita *et al.* (1967), Liebmann (1972), and Lythgo (1980).

(Figure 9.23). Specular reflectors of this kind can be tuned to reflect particular wavebands (by making the thickness of the guanine crystals and their spacing appropriate); those of deep-sea sharks reflect best at the wavelength best transmitted by deep water.

This kind of tapetum is rare in teleosts; more usually (as in many catfishes or in the garfish, which has an orange eye-shine) the reflector lies in the retina itself, and consists of spherical reflecting bodies within the pigment epithelial cells where they extend between the rods and cones. Such retinal tapeta are quite different to the specular reflectors of sharks, for they reflect diffusely, so that a narrow beam of light falling on the retina will be reflected through a surrounding hemisphere.

The function of both kinds of tapeta has not yet been satisfactorily established. In sharks it was found that the rods were only half the length of those in corresponding teleosts (without tapeta), and contained only half the visual pigment. So the tapeta here give approximately the same sensitivity as if the eye did not have a tapetum. It is possible that noise due to spontaneous breakdown of visual pigment may limit the sensitivity of the eye, and if this is so, then specular tapeta may operate to improve the signal to noise ratio of the eye, for with less pigment, there will be less noise. In the case of retinal tapeta, this argument does not apply, and Nicol and his colleagues (1973) suggest that such tapeta may increase sensitivity at the price of loss of acuity in turbid waters.

One disadvantage of a reflecting tapetum is that it makes the eye conspicuous when it is illuminated, and this is avoided in many, but not all, sharks by the migration of pigment to occlude the tapetum, a process taking an hour or more which is a direct response to illumination. A similar direct response to light is shown by the visual cells in many teleost retinae which move within the retina so that rods are masked in bright light to prevent them being bleached.

9.3.5 The receptors and visual pigments

Rods and cones in most fish retinae are as well-differentiated as they are in the mammalian retina (Figure 9.15) though in elasmobranchs and lampreys, they are harder to distinguish morphologically. As we should expect, rods (which contain much more visual pigment than cones and are thus much more sensitive) are found in all fish retinae, but cones (which with less visual pigment, can only operate at higher light intensities) are absent from the retinae of fishes living in dimly-lit environments like the deep sea. Both receptor types contain visual pigments made up of a protein (opsin) linked to an aldehyde of Vitamin A_1 (retinal) or Vitamin A_2 (dehydroretinal). The pigments with retinal are sometimes called rhodopsins, those with dehydroretinal, porphyropsins (these absorb at longer wavelengths). Depending on the aldehyde, and also on the opsin and on the link between them, the visual pigment absorption curves have different maxima so that receptors containing these different pigments are most sensitive (i.e. capture most photons) at different wavelengths. An elegant demonstration of this was given by Tomita and his colleagues (1967) who recorded from single goldfish cones illuminated by light of different wavelengths (Figure 9.15). Their results are in excellent agreement with the maximal absorption of cone pigments measured *in situ* by microspectrophotometry (Liebman, 1972). Most measurements of the wavelength at which visual pigments absorb maximally (λ_{max}) have been made by extracting the pigment from the retina with weak detergent solutions, such as digitonin, and because the sensitive rods contain much more pigment than the cones, the extracted pigments are those from rods. In different fishes λ_{max} of the extracted pigments differs, and it was early realized that such rod pigments might be matched to the wavelength of light best transmitted by the water in which the fish live, thus giving the fish the most sensitively adapted eyes. Two groups of workers independently suggested in 1936 that this idea implied that fishes living in the deep ocean should have rod pigments with λ_{max} shifted towards the blue, at around 475 nm—light of this wavelength remains from daylight after it has been filtered by the upper layers of the ocean. Later independent measurements showed that the rod pigment was indeed most sensitive at around 475 nm

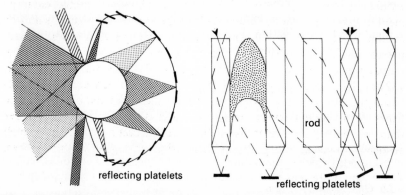

Figure 9.16 Reflecting tapeta. Left: orientation of reflecting platelets in *Squalus* tapetum. Right: Paths of light rays channelled by rods and reflected by platelets in sturgeon tapetum. Pigment cell black. After Denton and Nicol (1964), and Nicol (1969).

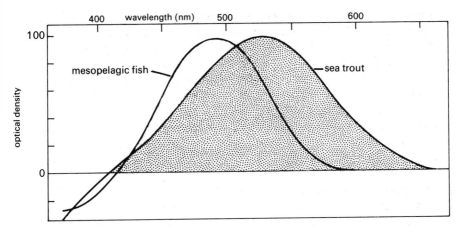

Figure 9.17 Rod pigments matched to the environment. Stippled, absorption curve for visual pigment extract of shallow water fish (sea trout, compared with that from a mesopelagic myctophid (*Diaphus rafinesqui*). After Denton and Warren (1957).

(Figure 9.17; see also Figure 9.24). Interestingly, in many teleosts, like the freshwater rudd, there are two rod pigments, one based on retinal, the other on dehydroretinal, with different λ_{max} values, and these change in relative amount during the year.

Where fishes migrate into waters of different spectral quality, as do eels and salmon, their rod pigments change. In the sea, coho and pink salmon have predominantly a retinal-based pigment (λ_{max} 503 nm), but as they begin to move into fresh water, this is gradually replaced by a dehydro-retinal pigment (λ_{max} 527 nm) more suited to the redder spectral environment of fresh water. In the freshwater eel, the pair of rod pigments absorb maximally at 501 nm and 523 nm, but when the eyes enlarge and the eel passes downstream to the sea, the 523 pigment disappears, and a new retinal pigment (with a new opsin) appears, with maximal absorption at 482 nm. The overall sensitivity of the mixture of the new and old retinal-based pigments lies between 482 and 501 nm and is like that of mesopelagic fishes. As we should expect, since fresh water is usually somewhat redder than sea water, freshwater fishes have dehydroretinal pigments which absorb at longer wavelengths than their retinal analogues in marine fishes.

9.3.6 Offset visual pigments

Many shallow-water fishes have visual pigments that are *not* exactly matched to the wavelengths best transmitted by the water in which they

live. Lythgoe (1979) suggested that although most fishes often need to see
as far as possible through the water (i.e. to use visual pigments matched to
the background light), another requirement is to detect prey and predators
against the background: to detect contrast. Objects darker than the
background, or distant objects whether lighter or darker are best perceived
by visual pigments matched to the background, but this is not true for the
special (but common) case of bright objects seen at fairly close range in
shallow water. Such objects are illuminated (and reflect) light which has
passed through a short water path, and from which, therefore, less of the
total waveband has been absorbed by the water filter than the background
light, which has on average traversed a longer water path. Such an object
will appear brighter than the background at all wavelengths, but the
important point is that the *relative* amount by which it radiates more light
than the background will be greater the more the wavelength departs from
that best transmitted by the water. So its contrast is enhanced by viewing
it with a visual pigment whose maximal absorption is offset from the
wavelength best transmitted by the water.

At greater depths, the downwelling light illuminating the object will be
progressively more and more similar in waveband to that of the
background, so that the difference in radiance between the object and its
background becomes the same at all wavelengths, and a matched pigment
is superior; at long distances, the light radiated by the object passes
through a long water path before it reaches the eye, and the same applies.
So it is only shallow-water fishes interested in nearby bright objects that we
should expect to have offset visual pigments. McFarland and Muntz
(1975) studied the visual pigments of two pelagic predatory fishes which
have different hunting strategies, the skipjack tuna and the dolphin fish
(*Coryphaena*). Skipjack maintain an elevated body temperature (Chapter

Katsuwonus

λ_{max} 483 nm

Coryphaena

λ_{max} 521, 499 and 469 nm

Figure 9.18 Hunting strategies and visual pigments. The dolphin (*Coryphaena*) hunts by
striking horizontally, the skipjack (*Katsuwonus*) by rushing upwards (see text). After Lythgoe
(1980).

3) and can thus range down into colder water; they hunt by rushing upwards at fish silhouetted against the surface (Figure 9.18). Consistent with Lythgoe's hypothesis that the visual pigment in skipjack should match the background light (since they are hunting *dark* objects), extracts have a single pigment which at λ_{max} of 483 nm is matched to the background. Dolphin fish hunt in an entirely different way, being confined to the upper 15 m of the water column, and striking horizontally at their prey. They have three extractable visual pigments (λ_{max} 521, 499, and 469 nm), and of these the 499 pigment is probably contained in the rods. The 521 pigment seems well-designed to detect bright objects (the flying fishes, crustaceans and garfishes on which the dolphin feeds), whilst the 499 pigment is probably matched to the background light. The function of the 469 pigment is not so clear, but may be offset for blue-green water rich in phytoplankton.

9.3.7 Colour vision

Single receptors are most sensitive, and will capture most photons, at the wavelength of maximal absorption of the visual pigment they contain, but they cannot distinguish between a dim light at this wavelength, and a more intense one at a different wavelength. This is why special precautions have to be taken in such experiments as that illustrated in Figure 9.16 to ensure that the quantal flux at different wavelengths is the same. To distinguish between light of different wavelengths (for colour vision) it is necessary to compare signals from at least two kinds of receptors that differ in spectral sensitivity, that is, in the visual pigments they contain. It seems not unlikely, as Lythgoe (1979) points out, that the roots of colour vision may lie in the mechanisms for contrast detection by offset visual pigments we have just seen. If pigments of different spectral sensitivity are best suited to detect the contrast of bright and dark objects, then the fish need only compare the signals from the different receptor types illuminated by the same part of the image to be able to distinguish colour.

Considering the bird-like brilliance of the colour patterns of many shallow-water teleosts, it is not surprising that they have colour vision. As we have seen (Figure 9.15) goldfish have three cone types (as we do), but red-sensitive cones are often absent from fishes living in turbid coastal waters, where there is very little red light in the environment. The different cones are often arrayed in superb mosaics, which have been determined by microspectrophotometry of retinae on microscope slides (Figure 9.15).

Red light is absent in the deep sea, apart from the special case of those fishes which shine red lights on their prey and whose red-sensitive rods contain a pigment with λ_{max} of 575 nm. Such fishes are operating on private wavelengths, undetectable by the blue-sensitive rods of their prey! Less is known of colour vision in other groups, but pelagic and reef sharks seem to have colour vision (Gruber, 1975).

9.3.8 Detection of light other than by the eyes

This brief account of some aspects of vision in fishes is scarcely more than a glance at a few of the fascinating adaptations fish have devised for seeing in water. Before turning to the way that many fish try to *prevent* themselves being seen, we should recollect that light is detected by some fish elsewhere than by the eyes. In lampreys for example, the lateral line receptors along the body are light-sensitive (perhaps because there the pigmentation of the skin is interrupted, and the effect is a direct one upon nerve fibres); in many fish there is a window above the pineal which is light-sensitive and has rod-like receptor cells.

9.4 Camouflage

9.4.1 Camouflage by reflection

Many fishes living in the upper layers of the ocean are silvery, because of silvery scales (herring) or underlying silvery layers (mackerel). The silveriness results from light reflection by organized stacks of thin guanine crystals, separated by sheets of cytoplasm, as in the tapetal reflectors. Such layers of material of alternating high and low refractive index operate as very efficient reflectors. Since the wavelength reflected depends upon the optical thickness (refractive index × thickness) of the layers, the highest reflectivity being when the optical thickness is $1/4\,\lambda$, such reflectors can be 'tuned' to reflect particular wavelengths.

Because the polar diagram of light in the sea (Figure 9.19) is symmetrical except just at the surface, it is possible for fishes to camouflage themselves by reflecting light (Denton, 1970). Consider a flat mirror vertical in the water. An observer (a predatory fish) looking at the mirror obliquely from A or B (Figure 9.19) will see light reflected from the mirror, but will be unable to distinguish this from light which would have reached its eye if the mirror were not there, and hence will be unable to detect the mirror. Although fish are not flat-sided, they can orient the reflecting platelets vertically (Figure 9.19) and achieve the same result as if they were flat. The

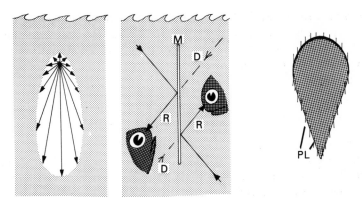

Figure 9.19   Light in the sea and camouflage with silvery surfaces. Left: distribution of radiance in the sea. The length of the arrows indicates relative radiance in a given direction from their origins. This polar diagram of radiance distribution remains constant with depth, although light intensity changes. Middle: predatory fishes looking at a perfect mirror (M) cannot distinguish between reflected rays (R) and direct (D) rays, hence the mirror is invisible. Right: a herring has reflecting platelets (PL) on its sides inclined slightly upwards, thus compensating for less than perfect reflection by reflecting light slightly brighter than if oriented vertically as in middle diagram. Note narrow keel and dark pigment dorsally. After Denton (1963).

bottom of the fish will be most difficult to camouflage in this way, since if looked at directly from below, it will be silhouetted against downwelling light: two solutions are possible. Either the fish can make the ventral region compressed and knifelike (Figure 9.19) as do herrings and sprat, or it may generate its own light to mimic the downwelling light, using photophores.

9.4.2 Luminescence and photophores

Light organs of different kinds are an intriguing and important feature of many mesopelagic and deep sea fishes, and are even found on a number of fishes living in shallow water, such as the midshipman *Porichthys* (no freshwater fishes are known to be luminous). Samples of fishes collected off Bermuda and in the South Atlantic showed that some 70 % of all species captured had light organs, and systematic surveys have shown that around 10–15 % of all marine fish genera contain luminous species. Some fishes use lights as lures—barbels and fishing rods with luminous tips are found in several families (Figure 9.20), whilst others have light organs inside the mouth—or as headlights to illuminate their prey, but the majority use their photophores for signalling to other members of the same species, and for

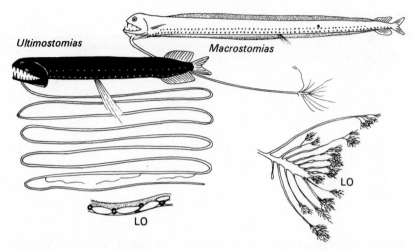

Figure 9.20   Stomiatoid barbels. *Ultimostomias* has light organs (LO) aligned along the barbel (below) whilst *Macrostomias* has luminous organs (LO) at the tip. Note ventrally directed camouflage photophores along body. After Beebe (1933).

camouflage. In the upper 1000 m of the oceans, downwelling light will silhouette fishes looked at from beneath, and it seems that many fish surmount this difficulty by shining light downwards to match natural daylight (Figure 9.21) and thus render themselves invisible. Since it is necessary also to match the background when viewed obliquely, it is hardly surprising that the most complex photophores and photophore arrays have been developed for ventral camouflage.

Fish photophores are fascinatingly varied, and their structure can only be touched on here (see Herring and Morin, 1978). There are two main types, one in which light is produced by special photocytes, and another where the fish employs symbiotic luminous bacteria, cultured in special sacs.

9.4.3 *Bacterial photophores*

Bacterial light organs are always either associated with the gut, and open to it, or are open to the sea; the three species of *Photobacterium* found in these light organs infect the chambers during larval life. They glow continuously, and so the fish can only 'switch off' bacterial photophores by masking them with a shutter or with chromatophores, or by rotating them into a black-lined pocket. The bacteria are present in colossal numbers in

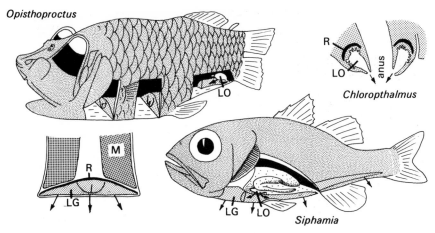

Figure 9.21 Bacterial light organs. Left: *Opisthoproctus* has a rectal light organ (LO) which transmits to a flattened light guide (LG) backed by a reflector (R) that emits light along the 'sole' of the fish, seen in transverse section below. M: myotomal muscles. Right: anal light organs (LO) of *Chloropthalmus* open from the rectum and are backed by reflectors (R). Bottom: the cardinal fish *Siphamia* has a light organ (LO) opening from the gut and illuminating a light guide (LG) which emits along the ventral surface. After Herring (1977), Somiya (1977) and Iwai (1971).

such photophores; in the spectacularly luminescent flashlight fish *Photoblepharon*, the light organs contain 10^{10} bacteria cm^{-3}! *Photoblepharon* lurks in reef caves in the tropics during the day, flashing intermittently (by blinking the shutter over its light organs); at night it emerges and hunts small copepods by the more or less continuous light it produces. Some other bacterial light organs like the perianal organ of *Chloropthalmus* (Figure 9.21) are very much dimmer than those of *Photoblepharon*—they were only discovered by a dark-adapted observer and probably are used as cues for schooling.

A number of fishes with bacterial light organs use them to illuminate the ventral surface, presumably to avoid being silhouetted by downwelling light, but externally show no special modifications. The light organ is a diverticulum of the gut, and is surrounded dorsally and laterally by a connective tissue reflecting layer. Light therefore emerges downwards, and is refracted by translucent ventral muscles before passing out of the ventral region of the fish (Figure 9.21). The most striking example is *Opisthoproctus* (Figure 9.21) which has a bacterial light organ near the anus formed from a gut diverticulum. It is enclosed in black epithelium except anteriorly, where it shines into a long ventral hyaline light guide

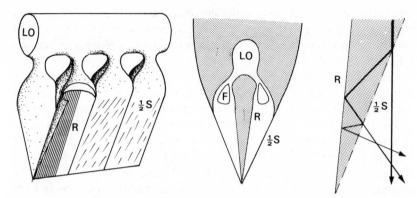

Figure 9.22 The organization of the ventral photophores of *Argyropelecus*. Left: stereogram of system showing dorsal light pipe (LO) connected to ventral reflecting chambers (R) backed with a curved reflector and half-silvered on the external surface ($\frac{1}{2}$S). Middle: transverse section through ventral part of body showing light organ (LO), filters (F), and reflecting chambers (R) with half-silvered outer surface ($\frac{1}{2}$S). Right: reflections of rays emitted from light organ into reflecting chamber giving rise to appropriate distribution of radiance for ventral camouflage. After Denton (1963).

surrounded dorsally by reflecting platelets. The bottom of the fish is completely flat, and light emerges evenly over the whole of this flattened sole. It seems that this remarkable system must be used for ventral camouflage in the upper mesopelagic zone where *Opisthoproctus* lives, but we do not know how the system is tuned.

9.4.4 Photophores with intrinsic photocytes

Perhaps because of problems of infection and maintenance, fishes with bacterial light organs have four at most, and usually only one or two, so that the same organ may serve several purposes, as in the flashlight fishes, where the organs are used to illuminate their prey as well as for schooling and sexual communication. Fishes with photophores where the light is generated intrinsically, on the other hand, often have many organs, and different ones are used for different purposes. Thus in the lantern fishes (Myctophidae), whilst ventral photophore series seem to be used for camouflage, the lateral photophores (differently patterned in the two sexes) are evidently used for intra-specific signalling. In some fishes, different photophores produce light of different colours.

Such photophores are often somewhat similar to eyes, for the photocytes may be backed by a reflecting layer, and capped with a lens. They are richly innervated, and are certainly under nervous control. In *Porichthys*, which has received most attention since it is accessible for experiment in good condition, the transmitter is adrenalin or noradrenalin, whereas in hatchet-fishes it is epinephrin.

The most complicated photophores so far studied are those of the hatchetfish, where they form groups of very conspicuous ventrally directed tubes along the lower part of the fish (Figure 9.13). In each group, light is produced in a dorsal chamber, lined with black pigment except for a series of small windows into the photophores (Figure 9.22). This light passes through a colour filter transmitting at 485 nm and then enters the wedge-shaped photophore. This is lined with a reflective guanine layer, and the external surface is covered with a half-silvered mirror, again made from guanine crystals. The result of this complicated design is that light entering the photophore from the dorsal chamber will emerge in a particular pattern (Figure 9.22), different in intensity at different angles. Because the reflecting surfaces on the inner borders of the photophores are curved, each photophore will distribute its light in a wide arc. Denton and his colleagues

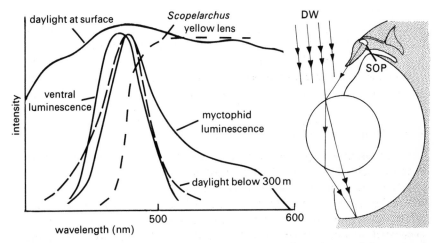

Figure 9.23 Ventral illumination camouflage and yellow lenses. Left: spectral characteristics of surface daylight and daylight at 300 m (stippled) compared with light emitted by *Argyropelecus* ventral photophores (hatched), and by myctophid photophores. Note close match of light emitted by *Argyropelecus* with downwelling daylight. The dashed curve shows transmission by the yellow lens of *Scopelarchus*, showing that the myctophid (but not the *Argyropelecus*) photophores will appear brighter than the background when viewed with such a yellow lens. Right: comparison of intensity of light emitted by supraorbital photophore (SOP) with downwelling daylight, in the myctophid *Tarleton-beania*. After Denton (unpublished), Lawry (1974) and Muntz (1976).

(1972) set up hatchetfish in a chamber in which they could be rotated and the light emitted measured with a photomultiplier. With this simple apparatus, they showed that the angular distribution of the light emitted by the fish was remarkably close to that of ambient light in the sea making it virtually certain that these photophores indeed function to make the fish hard to see from below.

In addition to matching the wavelength and angular distribution of the light emitted to that of the ambient light, the intensity of the light emitted must also match the ambient light. Hatchet fishes and many myctophids have small photophores so arranged as to shine into the eye, and these may act as reference sources to allow matching of ambient light. By adjusting the brightness of the ocular photophore (assumed to be functionally coupled to that of the ventral photophores) until that part of the retina illuminated by the photophore receives the same amount of light as that illuminated by downwelling light, the fish would be able to match its background. In the myctophid *Tarletonbeania* (Figure 9.23), Lawry (1974) has shown how such a mechanism could work, and has also found that low-intensity dorsal illumination of the fish in an aquarium causes it to illuminate its ventral photophores. However, direct experimental proof for this attractive idea is still lacking, and we have to admit that some fishes which seem to have ventral photophores arranged for camouflage do not appear to have any such matching mechanism.

The light output from hatchetfish is about equivalent to the intensity of downwelling light around 600 m (where the fish are found during the day); much nearer the surface fish cannot produce sufficient light for ventral camouflage and rely instead on transparency or on special morphological modifications.

9.4.5 Yellow lenses

The adaptations of fishes are so remarkable that it will come as no surprise to find that a few fishes have devised a means of cracking the ventral photophore camouflage system, using filtering lenses. The ventral light emission spectra of *Opisthoproctus* and *Argyropelecus* (Figure 9.23) are extraordinarily close to that of downwelling daylight, and yellow lenses (which *Argyropelecus* itself possesses) are of no help in detecting such fishes from below. But the emission spectrum of myctophid photophores is much broader; examined with an upward-looking eye with a yellow lens they will appear brighter than their background. Muntz (1976) has calculated that their camouflage could be pierced at a distance of some 16 m in the clear mesopelagic zone. It seems, however, that yellow lenses (as in hatchetfishes, and the predatory *Scopelarchus*) may more usually be used to increase the visibility of lateral photophores.

9.4.6 Red headlight fishes

Finally, an extraordinary special case of the use of photophores to circumvent the camouflage of prey is provided by fish which emit *red* light, and possess visual pigments absorbing red light. *Malacosteus* and *Pachystomias* have large red-emitting headlight photophores underneath the eye and their retinal pigments absorb at around 575 nm (Figure 9.24) so that they are able to perceive red light. Most deep sea fish have pigments with an absorption maximum around 450–490 nm, and so are unable to perceive red light. Such fishes as *Malacosteus* are able to illuminate their prey with light of a wavelength that the prey cannot detect, and can readily observe the common red and brown animals of the middle depths which

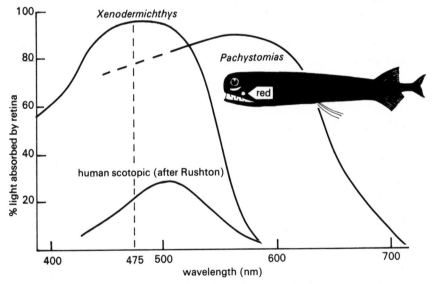

Figure 9.24   Private wavelength of *Pachystomias*. Curves showing light absorbed by single pass through retina. The dotted line is wavelength maximally transmitted by seawater at 300 m. Note that the alepocephalid *Xenodermichthys* absorbs maximally at the dotted line; human rods are suited to dim light (moonlight is similar but less intense than daylight, Figure 9.23), whereas *Pachystomias* absorbs maximally much further to the red end of the spectrum. *Pachystomias* has a red-reflecting tapetum and thus maximum absorption of the intact eye will be higher than that of the retina alone. After Denton (unpublished).

are not seen when illuminated with the more common blue-emitting photophores, like those on the head of the lantern fish *Diaphus*. Transmission of red light is not as great in the sea as blue light, and *Malacosteus* has a red-reflecting tapetum and increased pigment density in the retina to make up for the inevitable loss of sensitivity.

9.5 Olfaction, taste and pheromones

Many fishes are remarkably sensitive to olfactory stimuli, and correspondingly, have large olfactory lobes (Chapter 10). Experiments on lemon sharks (*Negaprion*) for example, where electrical activity was monitored by implanted electrodes in forebrain and medulla, whilst chemicals were introduced into the tunnel respirometer where the fish were kept, have shown thresholds of 10^{-9} M for a series of amino acids. Conditioning experiments with eels have shown that these fish can detect β-phenyl ethyl alcohol at the very low concentration of $1-3 \times 10^{-18}$ M, corresponding to 770 molecules ml^{-1} of water, which is equivalent to a single alcohol molecule only in the olfactory chamber! Direct records from the olfactory nerves of salmon when the olfactory epithelium was perfused with water from different sources showed that they were most sensitive to their home river water, but indifferent to the home waters of others of the same species.

The olfactory detector is arranged in essentially the same way in all fishes and consists of a chamber lined with folded olfactory epithelium, through which water usually flows unidirectionally. Flow through the olfactory sac may result from the forward motion of the fish, from the action of cilia lining the sac, or by active pumping, caused by respiratory movements deforming accessory sacs (as in *Polypterus*). In lampreys, water is inhaled and exhaled into the nasal sac by the action of the respiratory musculature. Commonly, the external nares lie close together on top of the head, but bizarre modifications like those of *Polypterus* or some coral fishes are also found, where the nares may lie on top of long tubes.

The receptors of the olfactory epithelium seem to be of two kinds, but nothing is yet known of any functional difference between the two. Unlike other vertebrate receptors, the olfactory receptors are primary sensory cells, sending their own axons to the brain via the olfactory nerves. These are often conveniently long, so that in long-snouted fishes (like the gar, *Lepisosteus*), they offer an unusually suitable source of small, uniform-diameter myelinated axons.

With such sensitive detectors for dissolved substances, it is not surprising that many fishes use olfactory cues to detect food, as do reef sharks like the blacktip and whitetip, which approach bait from downstream following an olfactory trail. Olfactory cues are also employed by fishes in their social behaviour. Female *Lebistes* respond to water in which males were previously kept by assuming the copulatory posture, and isolated males of *Bathygobius* begin their courtship behaviour 5–10 seconds after addition of water from an aquarium containing a gravid female. An ingenious series of experiments by von Frisch and his students showed that minnows (*Phoxinus*) could not only distinguish other fishes of different genera using olfactory cues, but could also distinguish between two different fishes of the same species. Freshwater Ostariophysi (and a few marine species) have special club cells in the skin secreting an alarm substance (*Schreckstoff*) if the skin is damaged that causes immediate changes in schooling behaviour. The original work on *Phoxinus* by von Frisch (1938) has been extended to other species, but the pheromone has only been found in Ostariophysi, and one cichlid. Other fishes exhibit alarm reactions to substances produced by predators (salmon have been shown to be sensitive to the odours of bear and dog paws as well as to that of the human hand!), and minnows are sensitive to water in which pike have been kept.

All of these responses have been shown to be olfactory (they are abolished by section of the olfactory nerves), but fishes are also equipped with large numbers of solitary chemoreceptors and tastebuds (Figure

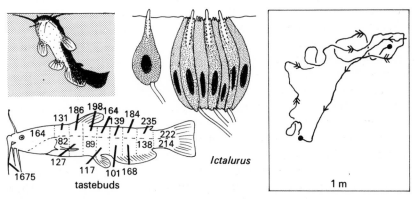

Figure 9.25 Taste in the catfish *Ictalurus*. Upper left: catfish seeking barbel contact with chemical in surface film. Lower left: distribution of tastebuds over body surface. Middle: single chemosensory cell, and tastebud. Right: paths of intact catfish (single arrows) towards chemical release in slowly flowing water, and of blinded catfish towards chemical release in still water (double arrows). After Bardach *et al.* (1967) and Bardach *et al.* (1969).

9.25), which in some specialized fishes are used in the same way as the olfactory system, for detection of distant food. Bardach and his colleagues (1967) showed that bullheads (*Ictalurus*) could detect food over 5 m away using the tastebuds of the barbels and body (Figure 9.25). Tastebuds contain secondary sensory cells, innervated by fibres in cranial nerves VII, IX and X which synapse at the bases of the sensory cells. Conditioning experiments to determine taste thresholds have only been carried out on a few fishes, for example, minnows can detect sucrose at 2×10^{-5} M and NaCl at 4×10^{-5} M, much more dilute solutions than we can ourselves detect.

Tastebuds are often found on special organs, such as barbels (Figure 9.25), or free fin rays which may be highly motile and used to probe the substrate for food. Mullets and goatfish have exceptionally mobile barbels which are equipped with numerous proprioceptive sensors so that the fish locates food and then can turn and snap it up.

CHAPTER TEN

THE CENTRAL NERVOUS SYSTEM

The central nervous system in fish is perhaps simpler than in other vertebrates, but it is nevertheless complex, and contains so many interconnected neurones that it poses a special problem for the student seeking a sensible short summary of present views on structure and function. The brain is conventionally divided into fore, mid and hind brain, and the structure and operation of each is separately considered. Certainly, because most texts deal with the brain of vertebrates in this way, such an arrangement makes it easier to compare the fish brain with that of other groups. But this approach is misleading, for it artificially simplifies the real interdependence of the different regions of the brain, often not only connected by tracts passing in one direction, but also by reciprocal connections. For example, in teleosts, the optic tectum projects to the isthmic nuclei in the caudal tegmentum, and the isthmic nuclei themselves project forwards to the tegmentum (Ito *et al.*, 1981). The older work on the anatomy of the fish brain was mainly based on methods which electively demonstrated cell bodies and fibre tracts, but great advances have been made in the past decade by the use of silver methods designed to show degeneration following section of fibre tracts or local lesions, and in particular, by methods such as the horse-radish peroxidase (HRP) technique, where cells and tracts are labelled by their uptake of HRP injected at specific sites. These methods have revolutionized knowledge of the interconnections of different brain regions and have revealed quite unsuspected projections, for example, projections of the electrosensory system into the forebrain in some teleosts. Ideally, therefore, it would be best to consider brain anatomy in terms of systems, rather than regions, and in this brief account we have compromised between a conventional account of the brain regions, and a survey of various systems.

BIOLOGY OF FISHES

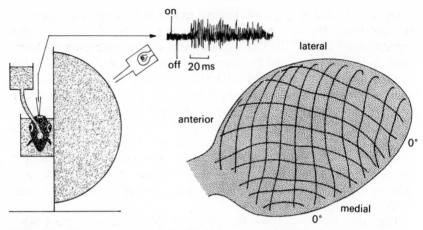

Figure 10.1   Somatotopic map of visual field in optic tectum. Left: experimental arrangement showing light spot illuminating surface of water-filled sphere centred on eye. Upper: extracellular responses from units activated by brief light flash (on-off shows stimulus artefacts). Right: map of responses in right optic tectum after stimulation of left visual field. After Schwassmann and Kruger (1965).

The general function of the brain is to receive and analyse sensory input, and to integrate different sensory inputs in order to initiate the appropriate motor outputs, but only in some rather stereotyped instances (as for example the giant Mauthner cell system, or the curious mesencephalic nucleus in dogfish) is much known of the steps in these processes. If we were ourselves designing circuitry for an analogous instrument to perform this kind of function, we should probably arrange that the first step in the spatial analysis of visual, lateral line, and electroreceptive input would be the same, that is, we should arrange a spatial representation or map of the input which could then be scanned and processed to some degree before feeding into more complex integrative circuitry. So far as is known, this does happen in the fish brain, where somatotopic representation of visual and electroreceptive input is well established (Figure 10.1). Probably the acoustico-lateralis system is also somatotopically represented, though this remains to be demonstrated.

The corollary of this sort of input representation is that we might expect to find similar circuits built from particular neurone types to process such input maps (and indeed, tectal and cerebellar circuits are somewhat similar) but the situation is complicated because in different sensory systems, different features of the map may be processed simultaneously. For example, in the visual system the map contains projections of edge detectors involved in shape recognition, as well as movement detectors. In the electrosensory system of gymnotids, something of a map of the aquatic space around the fish is represented in the cerebellum (Bullock, 1982), but this is processed by a complicated array of cells, some of which respond to objects which lie at particular points in the electric field generated by the fish, others to movement of objects in the field. There certainly are special and complex neurone circuits in different parts of the brain (the cerebellum is a particularly well-studied example anatomically), but only some arrangements of neurones can now be identified as low pass filters or as negative feedback circuits for example. Once more is known of the details of interconnection of brain neurones and of their excitatory and inhibitory effects on other neurones, more fruitful comparisons with artificial circuits will be possible.

The spinal cord is simpler than the brain, and although in most fishes there is probably segmental proprioceptive input involved in locomotion (see Chapter 3), here we are concerned with segmental and (relatively) simple patterns of neurones involved in the generation of rhythmic swimming rather than with the complex problems of input analysis.

The general plan of the central nervous system is the same in all fishes, although there are very considerable differences in the relative development of the different regions (Figure 10.2). Olfactory input passes to the

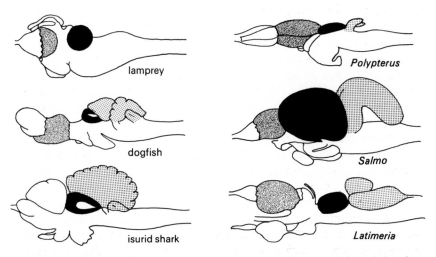

Figure 10.2 Brains of different fishes showing relative sizes of regions. Telencephalon, fine stipple; cerebellum, coarse stipple, optic tectum, black. After Ebbesson and Northcutt (1976), Roberts (1981) and Kremers (1981).

forebrain (prosencephalon), visual input to visual centres in the mid-brain (mesencephalon), and acoustico-lateralis input to the hindbrain (rhombencephalon). The brain can be regarded as an enlarged anterior portion of the spinal cord with hypertrophied centres associated with the development of input from the special sense organs. Figure 10.2 shows the arrangement in different fish groups; the three main regions of the brain are often sub-divided in the following way:

Forebrain (prosencephalon)	Olfactory lobes or
	cerebral hemispheres (telencephalon)
	Between-brain (diencephalon)
Midbrain (mesencephalon)	Optic lobes
Hindbrain (rhombencephalon)	Cerebellum (metencephalon)
	Medulla oblongata (myelencephalon)

The medulla oblongata tapers into the spinal cord, with a narrow central canal, the posterior continuation of the brain ventricular system. A schematic diagram of the regions of the brain (Figure 10.3) will help the reader find his way about the next sections.

10.1 Spinal cord

As in all vertebrates, the central nervous system arises by the inrolling and fusion of neural folds, giving rise initially (and usually permanently) to a more or less circular-section spinal cord, where a central grey area is surrounded by the fibre tracts of the white. In fish above cyclostomes, the central grey is more or less cruciform in section (Figure 10.4) so that dorsal and ventral horns like those of other vertebrates are found, but in cyclostomes (though not in amphioxus) the cord becomes flattened and ribbon-like during ontogeny, and these horns are not evident. Within the cord, the cell bodies of spinal neurones lie in the grey, (with a few exceptions in sharks and cyclostomes) and these are arranged in segmental patterns. Dorsally in the cord there are neurones and fibre tracts forming the somatic sensory and visceral sensory divisions; ventrally, the visceromotor and somatic motor systems are composed of the ventro-lateral masses of motoneurones in the grey matter and the ascending and descending fibres which form the large ventral-fibre tracts. Neuroglial cells in the cord are concentrated around the central canal (through which passes the curious Reissner's fibre secreted by a group of ependymal cells in the diencephalon); in cyclostomes glial elements are simpler than in other vertebrates, and the nerve fibres of the cord do not have myelin sheaths.

The neurone types of the spinal cord in fish have not yet been certainly classified into functional categories, for example, motoneurones have only been identified physiologically in lampreys and dogfish, and this is particularly unfortunate, for crude experiments clearly show that the

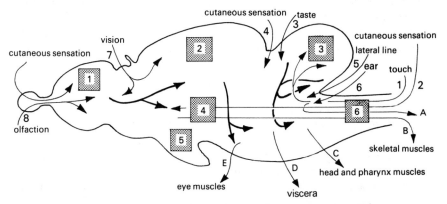

Figure 10.3 Generalized fish brain showing functions of different regions. Stippled rectangles: regions of general integration. 1, learning, appetitive behaviour, attention; 2, sensory coordination, motor integration; 3, postural control and autonomic regulation; 4, alerting mechanisms; 5, homeostatic and appetitive coordination; 6, motor coordination. Efferents: A, reticulospinal tract, B, Vestibulospinal tract; C and D, cranial nerves 5, 7, 9, 10, 11; E, cranial nerves 3, 4, 6. Afferents: 1, spinocerebellar tract; 2, spinoreticular tract; 3, cranial nerves 7, 9, 10; 4, cranial nerves 0, 5, 7, 9, 10; 5, cranial nerves 7, 9, 10; 6, nerve 8; 7, nerve 2; 8, nerve 1. After Laming (1981).

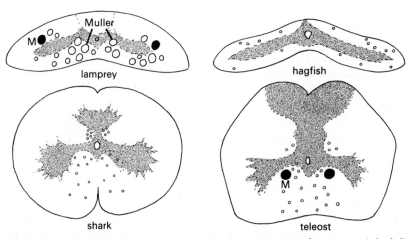

Figure 10.4 Spinal cord. Transverse sections showing arrangement of grey matter (stippled). Note large Mauthner axons (M) in teleost and lamprey, and the Muller axons of similar size in the lamprey (Muller). After Rovainen *et al.* (1973) and Niewenhuys (1964).

segmental patterns of spinal neurones are of prime importance in the way in which the spinal cord operates to generate the rhythmic swimming movements.

10.1.1 Spinal swimming

In sharks, destruction of the brain does not result in immediate paralysis of the body, as it does in cyclostomes and teleosts; 'spinal' sharks appropriately set up so that they are respired and supported off the bottom continue a stereotyped slow swimming pattern. In spinal dogfish, tail beat frequency is around $35 \, \text{min}^{-1}$, about the same as that of the minimum cruising speed sufficing to generate sufficient dynamic lift to maintain them off the bottom (Chapter 4). Stimulation of the body, or of localized regions of the cord, in such spinal fish modifies this slow swimming pattern, for example altering amplitude and frequency or by bringing the fast myotomal motor system into play. Clearly the spinal cord must contain fixed segmented neurone patterns, and intersegmental connections which control the serial myotomal and fin movements seen during swimming; the localized stimulation experiments indicate that the system is 'hard-wired' in such a way that separate circuits exist for each type of movement.

In cyclostomes and teleosts, remarkably enough, although recently spinalized fish are totally inactive, after some days chronic spinal fish show a similar regular spinal swimming pattern to that of sharks. Several possible explanations of this delayed response are possible (including neuronal growth and the establishment of new connections, or the substitution of spinal neurones maintaining a general level of excitation.) Thus not only in sharks is rhythmic locomotor activity an inherent property of the spinal circuits. In the non-swimming fish, descending fibres from the brain must inhibit segmental motor output.

In lampreys, recent studies of the isolated cord (Poon, 1980; Cohen and Wallen, 1980) have shown that rhythmic motor activity (similar to that seen in the swimming animal) is generated when L-DOPA or D-glutamate are added to the experimental bath. Figure 10.5 shows the experimental setup and records obtained when D-glutamate was added. Patterned motor output of this kind evoked by adding acidic amino acids and L-DOPA (at low concentrations) shows the same phase-lags between segments as in normal swimming, and is abolished by addition of known glutamate inhibitors. This suggests that in the intact lamprey, swimming is evoked by descending pathways with glutaminergic synapse with the local segmental circuits of the cord.

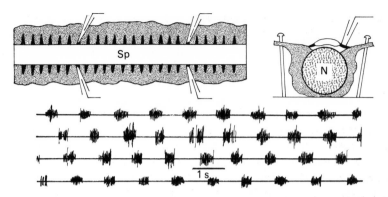

Figure 10.5 Rhythmic motor activity in isolated spinal cord. Above: experimental arrangement of lamprey spinal cord isolated to record from ventral roots (black) with suction electrodes. N: notochord. Below: simultaneous records from four suction electrodes on different right (upper pair) and left (lower pair) ventral roots showing rhythmic motor output in preparation paralyzed with curare and bathed in a solution containing 0.6 mM D-glutamate. After Cohen and Wallén (1980).

In spinal dogfish, a minimum of eight intact cord segments are needed for spinal swimming; 'fictive' swimming in the lamprey cord requires a minimum of four intact segments. The relatively small numbers of elements involved (there are about 500 neurones in each half-segment of the lamprey cord, and about 100 motor axons in a ventral root), and the transparency of the flattened cord which enables neurones to be seen and impaled in the living cord, obviously makes this a particularly attractive system for understanding the organization of the central network generating swimming movements (earlier work is summarized by Rovainen, 1979).

The anatomical basis for pattern-generating circuitry in the fish spinal cord is not known. As yet, correlations have not been established between the different cell types recognized functionally, and those seen histologically. Physiological investigations on dogfish (Roberts, 1981a) have shown that, for example, some motoneurones (identified by anti-dromic stimulation via the ventral roots) are spontaneously active, and fire in phase with the swimming rhythm, but the great majority of moto-neurones, like some of the interneurones, were usually found to be silent. Three classes of interneurones were recognized: the silent group, a group firing rhythmically correlated with the motor output, and a group discharging spontaneously and at high frequency unrelated to the motor output. How these differently-behaving neurones interact in normal

swimming remains a puzzle, but these studies may allow the elementary
schemes of Figure 10.6 to eventually become transformed into real circuit
diagrams. Other interesting questions raised are, for example, the
anatomical basis for the phase lags between different segments, involving
delays of some 20 ms between one ventral root outflow and the next, and
the differences between red and white motoneurones supplying the two
motor systems of the myotomes (Chapter 3).

One approach to the functions of the different neurones in the cord is to
correlate the anatomy of the simpler spinal cord of young embryos with
the gradual development of locomotor behaviour. Whiting (1948) has
examined the cord of the lamprey pro-ammocoete (in some respects the
most generalized basal vertebrate). The relation of the important
descending Mauthner fibre to the motoneurones of two types suggests that
one is likely to innervate white fast fibres, the other the slow red fibres.
There is a larval dorsal sensory system, the Rohon-Beard system,
providing segmental input to the system. Experiments on *Xenopus*
embryos, which also have a conspicuous Rohon-Beard system and rather
similar circuitry to the proammocoete, suggest that phase lags between
segments occur at this stage in development (Roberts and Clerk, 1980);
obviously the simple circuitry of Figure 10.6 is not sufficient for such
responses.

Descending fibres, forming a large ventral tract, are from the reticular

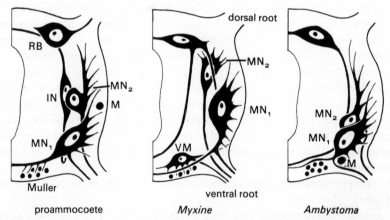

Figure 10.6 Cell types in the spinal cord of young chordates. Left: proammocoete larva of
lamprey. Middle: *Myxine* (cord shape transformed to that of proammocoete). Right: larval
Ambystoma. RB: Rohon-Beard cell, M: Mauthner axon, MN_1: primary motoneurone, MN_2:
secondary motoneurone. After Whiting (1948), Youngstrom (1940) and Bone (1963).

system on to which projects the motor outputs of the brain regions. There are also, at least in sharks, descending vestibular fibres from the large cell portion of the acoustic nucleus which descend near the central canal, forming Stieda's fasciculus. Perhaps these permit direct ipsilateral access of acoustic input to the segmental motor circuits of the cord in the absence of the specialized Mauthner system in adult sharks, but why this system is absent in sharks is not known.

As well as the descending fibres (which project to interneurones as well as directly to motoneurones) and the segmental elements such as moto-neurones, commissural interneurones and in many cases, giant dorsal cells, the spinal cord also contains visceromotor neurones and the ascending and descending fibres from dorsal root sensory input. Visceromotor neurones have not been identified in the fish cord (although they have been recognized histologically in amphioxus), but it seems that at least some of the large dorsal cells in the lamprey cord are viscerosensory (as some Rohon-Beard neurones probably are) since their axons pass out of dorsal roots and run to the viscera.

Compared with the spinal cords of higher vertebrates, which have received much more anatomical and experimental attention, the fish cord seems to be characterized by very many more dendro-dendritic connec-tions. For example, the motor cells of lampreys or dogfish receive relatively few synaptic inputs on the soma, but have enormous dendritic fields; Niewenhuys (1964) comments on the obvious reduction of the dendritic field of spinal neurones as one examines a series from 'lower' to 'higher' vertebrates. In the cord of the mammal, motoneurones possess recurrent collaterals, which, together with the Renshaw cell system, form a negative feedback loop. Supraspinal convergence on Renshaw cells perhaps allows recurrent inhibition to serve as a variable gain regulator at motoneurone level (Hultborn et al., 1979). As yet, Renshaw cells and recurrent collaterals from motoneurone axons have not been demonstrated in fishes. It may be that the damping of movements resulting from the density of the water means that fishes not only do not require sophisticated proprioceptive input, but also do not require such accuracy of timing between antagon-istic muscles as is needed in terrestrial forms, and it is for this reason that the complex feedback loops of the mammalian system have not appeared in fish.

A mysterious feature of the fish spinal cord is a system of neurosecretory cells at the hind end whose axons pass to a storage depot of neurosecretory material associated with a special blood plexus. This urophysis is in many respects similar to the hypothalamo-hypophysial system, and it seems to be concerned with salt regulation, producing a mixture of urotensins (which are known to have potent effects on the mammalian vascular system).

10.2 The brain

10.2.1 Brain size

The absolute size of the brain obviously differs in fish of different sizes (Figure 10.7), but brain/body weight ratios are similar for all groups of fish (and similar to those of amphibians and reptiles), with the exception of elasmobranchs, which have relatively much larger brains. Birds and mammals have much higher brain/body weight ratios than anamniotes, but some elasmobranchs at least approach the bird/mammal values, and have brain/body ratios as much as 400 % higher than other fishes. Why are there these differences? As Ebbesson and Northcutt (1976) point out in their review of anamniote brains, changes in brain size could be brought about by change in number of units; by change in the size and complexity of the units (e.g. by change in the size of the dendritic field of neurones); or by changes in the circuitry connecting the units. At present, only in certain

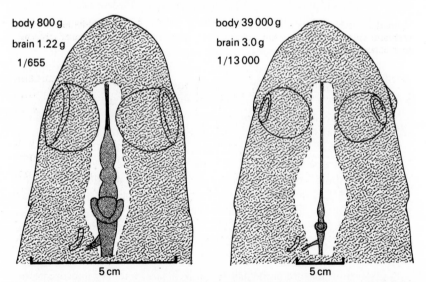

body 800 g

brain 1.22 g

1/655

body 39 000 g

brain 3.0 g

1/13 000

5 cm   5 cm

Figure 10.7   Changes in relative brain size during growth. Left: dorsal view of head of embryo from oviduct of the coelacanth *Latimeria*, right: similar view of adult. Note vast cranial cavity of adult compared with size of brain, and very large difference in brain/body wt. ratios between embryo and adult. After Anthony and Robineau (1976).

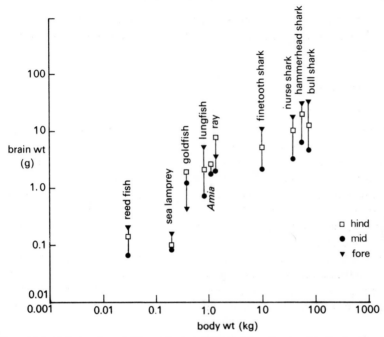

Figure 10.8   Relative sizes of different brain regions in fishes. Squares, hindbrain; triangles, forebrain; circles; midbrain. After Ebbesson and Northcutt (1976). Note relatively small brain of lamprey.

specific brain regions do we have a clear idea of which factor is dominant, as in the cerebellum of electrolocating fishes, where both neurone numbers and complexity have increased the cerebellar size as compared with 'ordinary' fishes (see Figure 10.13).

According to Szarski (1980), larger brains have fewer neurones per unit volume, but their processes (and neuroglial cells) occupy a greater volume than in smaller brains; as yet, DNA measurements on fish brains have not been undertaken to see if there is a relation between total cell number and brain size.

The relative size of different brain *regions* is very different in different groups (Figures 10.2, 10.8), and in different fish of the same group (Table 10.1).

10.2.2 Brain regions in elasmobranchs

The structure of the *olfactory bulbs* is basically similar in all vertebrates from cyclostomes to mammals. In each bulb, olfactory nerve fibres are

Table 10.1 Weights of brain regions in cartilaginous fishes as percentages of the total brain weight (from Northcutt, 1978)

Species	Olfactory bulbs	Telencephalon	Diencephalon	Mesencephalon	Cerebellum	Medulla oblongata
Hydrolagus colliei (chimaeroid)	7	31	6	14	21	21
Squalus acanthias (piked dogfish)	6	24	7	15	20	28
Etmopterus hillianus (deep-sea squaloid)	9	29	9	10	16	27
Mustelus canis (smooth dogfish)	14	32	7	11	16	20
Scyliorhinus retifer (chain dogfish)	14	35	7	12	15	17
Isurus oxyrinchus (mako shark)	3	37	6	18	19	17
Prionace glauca (blue shark)	3	45	8	13	13	18
Carcharhinus milberti (sand bar shark)	3	51	8	8	13	17
Sphyrna lewini (hammerhead shark)	7	52	4	6	19	12
Platyrhinoidis triseriata (thornback skate)	7	27	8	11	14	33
Raja eglanteria (clear nose skate)	3	32	7	13	17	28
Rhinobatos productus (guitar fish)	5	35	8	10	14	28
Myliobatis freminvilli (bullnose ray)	11	39	5	9	20	16
Dasyatis centroura (roughtail stingray)	8	47	5	8	18	14

gathered in glomeruli connected via interstitial cells to the olfactory tract fibres that run to the forebrain. While all cartilaginous fishes have large olfactory organs and bulbs, in certain species this system is extremely well developed. In the first five species in Table 10.1 (which are all of about the same size), the olfactory bulbs of the smooth dogfish and chain dogfish make up much more of the brain weight than do those of the other three forms. The same is true of the hammerhead shark in comparison with the mako and the blue shark; all three species attain much the same body size.

As in bony fishes, it is becoming clear that the *telencephalon* (forebrain) is not simply an olfactory centre. For instance, in the nurse shark (*Ginglymostoma*) and the torpedo-ray (*Torpedo ocellata*) each olfactory track leads to a relatively small centre in the lateral wall of the telencephalon. Indeed, new techniques for revealing experimentally selected neural pathways show that relatively large parts of the telencephalon in sharks and other fishes are concerned with vision and probably other modalities as well (Ebbesson and Northcutt, 1981). The efferent pathways are known only in a few species of sharks (and teleosts), in which descending fibres pass through the brain and reach as far as the first spinal segment. In sharks, in contrast to teleosts (and amphibians) most telencephalic efferent fibres cross at diencephalic levels and are then distributed to contralateral thalamic, tectal and reticular cell groups.

As already indicated, the *diencephalon* carries thalamic fibre tracts to and from the telencephalon. In fishes the hypothalamus bears a saccus vasculosus (of unknown function) and the hypophysis, which becomes associated with the anterior pituitary.

The *mesencephalon* consists of upper (optic tectum and torus semi-circularis) and lower (tegmental) regions. The layered tectum has a complex neuronal architecture; it receives and processes input from the optic nerves. Outflow in sharks (and in lampreys, teleosts and amphibians) consists of ascending tectal pathways that impinge on several pretectal and one or more thalamic nuclei. The descending outputs are formed of ipsilateral and contralateral tectoreticular pathways. The latter pathway usually reaches the anterior spinal region (Ebbesson and Northcutt, 1980). In sharks (and goldfish) there is also a projection from the cerebellum to the mesencephalon.

In most sharks the *eyes* are relatively small. The exceptions include the spiny dogfish (*Squalus acanthias*), the big-eyed thresher shark (*Alopias superciliosus*), the tiger-shark (*Galeocerdo cuvieri*) and fast-swimming isurids such as the porbeagle (*Lamna nasus*), mako (*Isurus oxyrinchus*) and great white shark (*Carcharodon carcharias*). The percentage figures for the weight of the mesencephalon are relatively high in *Squalus* and *Isurus* (Table 10.1). Chimaeras (*Hydrolagus*) and deep-sea sharks (e.g. *Etmopterus*) also have relatively large eyes and mesencephalon.

In cartilaginous fishes the *cerebellum*, which consists of a central corpus and lateral auricles, is relatively well developed. Among sharks it is best developed in the galeomorphs (e.g. *Isurus* and *Sphyrna*) and among rays in the myliobatiform species (e.g. *Myliobatis* and *Dasyatis*) (Table 10.1). In the forms with the largest cerebellum, the corpus is convoluted. It receives visual, auditory and trigeminal inputs, as well as ascending fibres from the spinal cord. Vestibular receptors project to the auricles, and lateral line input to the lateral line lobes.

Two prominent tracts of efferent fibres cross over to form a brachium conjunctivum, each half reaching the tegmentum, whence an ascending tract extends to the diencephalon. There is also a descending (contralateral) component of the brachium conjunctivum that is distributed to the medial reticular formation. An ipsilateral descending pathway extends backwards to the first spinal segment (see Ebbesson and Northcutt, 1980).

How the cerebellum may control movement has been recently investigated by Paul and Roberts (1979), see also Roberts (1981*b*). After ablation of the cerebellum in the dogfish (*Scyliorhinus*), reflex elevating

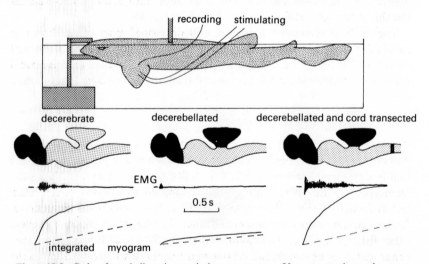

Figure 10.9 Role of cerebellum in regulating movements. Upper: experimental arrangement; middle: parts of brain removed (black); lower: muscle responses. EMG: electromyogram; below this, the integrated myogram. Note dotted line shows response of integrator in absence of EMG. After Paul and Roberts (1979).

movements of the pectoral fin are markedly changed (in that the magnitude of the response to electrical stimulation is much reduced and the threshold is raised, Figure 10.9). Roberts suggests that the cerebellum modulates the inhibitory drive to the spinal cord, so that during a particular movement the entire output of the cerebellum to the hind brain is shaped so as to selectively modify its influence on the cord. The cerebellum is seen not as being wholly involved in the pattern of a particular movement, but as controlling how much of the pattern, which originates elsewhere in the nervous system, should be expressed, acting via hind brain centres to modulate the 'gain' of reflex responses. Thus, different regions of the body are seen to be regulated by specific areas of the cerebellum, which holds a map of the body's motor activities. If this view is correct, the size and elaboration of the cerebellar cortex ought to be related to the amount of muscle to be controlled, the number of parts involved, and the complexity of the entire pattern of movement.

Certainly, the rapidly swimming isurid sharks have large cerebellums, as do stingrays and myliobatids (Table 10.1) which flap their pectoral fins; it is interesting that rays *do* have abundant proprioceptors in their fin muscles, and that these are especially numerous in stingrays. The problems of locomotor control must be greatest in the tail-less butterfly rays. We predict that they would have the largest cerebellums of all elasmobranchs, but this is not yet certain.

The *medulla oblongata* is also a well developed part of the brain in cartilaginous fishes (Table 10.1). As in other fishes, sensory centres account for most of its bulk. There are nuclei for the sensory fibres of the trigeminal nerve, the acoustico-lateralis nerves and the gustatory system (vagal lobe).

10.2.3 Brains of non-elasmobranchs

Telencephalon. The parts of the brain in other fishes may now be compared and contrasted with those in the cartilaginous fishes. Cyclostomes have a highly developed sense of smell. The bulbs and olfactory centres in the telencephalon are large (Figure 10.2) compared to the rest of the brain (Kleerekoper, 1969). The same is true of the lungfishes, *Polypterus*, *Latimeria*, sturgeons (*Acipenser*) and *Amia*. In teleosts, in contrast to these other fishes, the olfactory system, and no doubt its telencephalic centres, range widely in development. At one extreme there are groups such as the eels (Apodes) with highly developed olfactory organs (macrosmatic forms): at the other there are microsmatic species such as pike (*Esox*), or flying-fishes (Exocoetidae).

As in cartilaginous fishes, recently developed neuro-anatomical techniques have shown that the teleost telencephalon is not simply an olfactory centre—there are also well defined visual and auditory areas (Echteler and Saidel, 1981). In sharks only about 10% of the telencephalon is given to olfaction (Graeber, 1978). Clearly, there is much to be discovered. Besides its likely function as an integrator of sensory modalities, it is possible that much of the fish forebrain is the forerunner of the limbic system in mammals (Aronson, 1981).

The actinopterygian telencephalon is strikingly different to that of other vertebrates, for the roof is simply a thin sheet of ependyma, and lateral ventricles are absent. This condition of 'eversion' has made comparison with the forebrains of other vertebrates very difficult, but the recent results of Echteler and Saidel (using HRP injection and retrograde cell labelling) have shown that there are ascending thalamic tracts similar to those of terrestrial vertebrates, and that homologies can be traced between the dorsomedial region in the teleost and the corpus striatum, and between the two dorsolateral regions and the dorsal and medial pallium of terrestrial vertebrates. Despite differences in embryological development and in gross morphology between the forebrains of *Polypterus*, *Amia*, sturgeons and teleosts, and of other vertebrates, it seems that all are functionally organized in essentially the same way. The reader should bear these recent results in mind when looking at older work on the fish brain.

Most of what is known of the functions of the non-olfactory forebrain in fishes has come from studies of teleosts. Ablation (and electrical stimulation) has revealed much and raised many questions. First, total ablation does not seem to affect vision, response to sounds, taste, postural reflexes and gross motor skills. Species-typical feeding and aspects of learning (e.g. habituation, classical conditioning, escape learning) are also unaffected. Affected activities are olfaction, spontaneous activity, and startle reflexes (general category), aggression (threshold) and reproduction (species-typical category). Concerning learned behaviour, there seems to be considerable disruption of activities in tests involving punishment (or passive avoidance) and variable effects on appetitive learning rates (acquisition, somewhat slower; extinction, retarded; reversal learning, retarded; and delayed reward, disrupted)—see Flood and Overmier (1981).

Regarding the forebrain and spontaneous activity, it is necessary to distinguish between self-generated movements (spontaneity) and ability to respond to environmental change (initiation). Thus, telencephalon-ablated teleosts lack the latter but not the former (Savage, 1980; de Bruin, 1980).

Aggressive behaviour, as in defence of territory or in establishing hierarchies (shown by butting, biting, snapping, sound-making or threat display), is reduced after ablation of the forebrain in the wrasse, *Crenilabrus*, the bream, *Diplodus*, and the Siamese fighting-fish, *Betta* (de Bruin, 1980).

Behaviour related to reproduction is also changed after forebrain removal. For instance, the fighting-fish, *Betta splendens*, builds a floating nest of mucus-coated air bubbles, which are also produced in operated fish but not gathered into a cohesive nest. Forebrainless male sticklebacks reacted to responsive females by zig-zag dances but they were unable to lead them to a localized place in the tank (de Bruin, 1980). In *Macropodus opercularis*, the paradise fish, responses to the female, such as body displays, approach and chasing behaviour, which precede spawning, are diminished temporarily after ablation of the telencephalon. The effects of forebrain lesions on the incidence of spawning and egg cannibalism in male paradise fish have been studied by Davis *et al.* (1981). Lesions involved parts of the pallial cell masses (area dorsalis telencephali, *D*) and the subpallial cell masses (area ventralis telencephali, *V*). Spawning trials were then conducted with females known to be capable of spawning. The incidence of spawning was not significantly decreased following bilateral removal of area *D* and the posterior and dorsal parts of *V*, but full telencephalic ablation (with varying amounts of damage to the preoptic nucleus region) did lead to decreased spawning. However, it is clear that the part or parts of the forebrain involved in reproductive activities have still to be clearly located.

Suitable tests show that the telencephalon of teleosts is involved in the formation and maintenance of learned associations. For instance, fish with a damaged forebrain are inferior to sham-operated individuals in learning to avoid stimuli that are followed by an electric shock. Fishes lend themselves well to learning studies. The majority of teleosts have good colour vision and can be trained to discriminate colour and shape—indeed, fishes can respond to many of the training techniques used with mammals (Savage, 1980). For instance, fishes can be trained to respond to a conditional stimulus when the presentation of the unconditional stimulus is not contingent on the performance of the response. Colours, sounds, temperature changes, salinity changes, touch, smell and taste are among those that have been successfully used as conditional stimuli, while unconditioned responses have included general activity, electrical organ display in mormyrids, aggressive display (as in Siamese fighting-fish) and changes in respiration (Gleibman and Rozin,

1971). Under natural conditions, learning is important in territory recognition. Thus, salmon (*Salmo* and *Oncorhynchus*) learn the olfactory nature of their home stream, and must remember such odours until they return to spawn.

The visual centres. In fishes the optic tectum receives most of the nerve fibres from the eye. Pretectal nuclei also receive optic nerve fibres but their function is unknown. A number of axons from the tectum return to the eye, but the great majority of tectal efferent fibres sweep downwards and backwards to reticular nuclei in the midbrain and hindbrain that send tracts to descending motor columns. The latter exert a control on movement through motor cells in the spinal cord (see Guthrie, 1981).

In fishes with well-formed eyes the optic tectum consists of a cortex containing at least 6 distinct layers: an outermost marginal layer, an optic layer, a superficial grey and fibrous layer, a central grey layer, a central white layer, and a periventricular layer. The functions of cells intrinsic to these layers have been studied by Guthrie and Banks (1978) and Guthrie (1981). Their hypothesis is that intrinsic cells with long dendritic rami arranged at right angles to the tectal surface (these cells are much the most numerous), are related to functions dependent on the conservation of fine detail, especially as regards position (System 1). Cells with processes spread out in a plane more or less parallel to the tectal surface (System 2) are more likely to be involved in generalized functions such as the response to a moving object regardless of identity.

Each cell that is sensitive to visual stimuli will respond only to light coming from a limited part of the visual field of an eye, its receptive field (RF). The most detailed results obtained by Guthrie and Banks have come from the analysis of single-unit activity within the tectum by means of raster scanning (Figure 10.10). Impulses from a neurone appear as dots on the oscilloscope screen in register with the stimulus light-spot being swept across a screen in front of the eye (of a perch, *Perca fluviatilis*).

Receptive field (RF) types are either highly specialized with well defined boundaries (type 1 cells) or irregular and poorly defined (types 2 to 5). The former may well be involved in visual functions requiring precise information concerning position (location and contour): the latter respond more generally to features such as brightness, contrast, velocity and degree of novelty. Guthrie (1981), also relates these findings to the visual behaviour of perch. One of the most prominent kinds of intrinsic cells (termed pyramidal), which are set in the superficial grey and fibrous layers, is typical of teleosts (Vanegas, 1981). The apical dendritic shaft of a

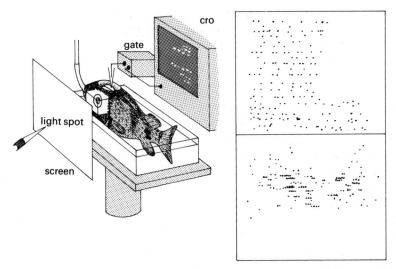

Figure 10.10 Responses of visual cells of *Perca* to different light stimuli. Left: experimental arrangement (note water-filled box around eye to minimize refractive errors). A light spot is scanned across the screen, and responses of tectal cells (right) are picked up, gated, and displayed on oscilloscope screen (cro). Right: upper, cell shows rostral inhibitory field, lower cell responds in a more localized pattern by bursts of activity. After Guthrie and Banks (1978).

pyramidal cell extends upwards into the superficial layer where it forms an espaliered dendritic tree with spines that make Grey type 1 synapses with boutons of marginal fibres (Figure 10.11). This is reminiscent of the Purkinje cells and parallel fibres in the cortex of the cerebellum (see Figure 10.14). Moreover, marginal fibres can be electrically excited to give action potentials rather like those of parallel fibres and the negative wave produced by depolarization of the apical dendritic tree of Purkinje cells. Indeed, marginal fibres originate in the torus longitudinalis whose main input comes from the valvula cerebelli in teleosts. Since the pyramidal cells receive retinotectal terminals, the marginal fibre system is a means whereby cerebellar influences meet the direct (monosynaptic) arrival of information from the retina.

Functions of the diencephalon. The diencephalon in fishes is much involved in vision: it receives retinal fibres and also an efferent tract from each optic tectum. In the nurse shark (*Ginglymostoma*), spinal cord and cerebellar fibres have been traced to the dorsal thalamus (Ebbesson, 1981). Here, we shall be simply concerned with hypothalamic mechanisms of feeding in

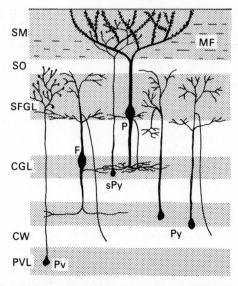

Figure 10.11   Cell types of the tectum. Some of the cells seen in the tectum impregnated by the Golgi technique. P, pyramidal; F, fusiform; Py, pyriform; SO, marginal layer; SFGL, superficial fibrous and grey layer; MF, marginal fibres; CGL, central grey layer; CW, central white layer; PVL, periventricular layer. Note some similarity with circuit arrangements of cerebellum (Figure 10.14). After Vanegas (1981).

fishes. In mammals hypothalamic areas regulate food intake. Moreover, in three teleosts, the sunfish (*Lepomis macrochirus*), a cichlid (*Tilapia macrocephala*) and carp (*Cyprinus carpio*), low-threshold and complete feeding responses can be evoked by electrical stimulation of areas near the lateral recess of the third ventricle into the inferior lobe of the hypo-

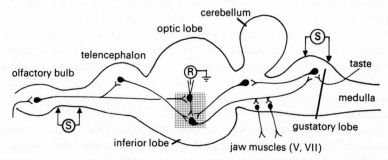

Figure 10.12   The hypothalamic feeding centre and its connections. S: stimulation, R: recording. After Demski (1981).

thalamus. One of the most consistent responses in the first two species was the snapping-up of gravel. Stimulation of the inferior lobe in sharks also results in feeding behaviour (Demski, 1981). Feeding responses may also be elicited by stimulation of telencephalic regions that are likely to be connected to the hypothalamic area. Hypothalamic lesions in goldfishes result in reduced feeding in individuals with damaged lateral or inferior lobes.

Figure 10.12 is Demski's (1981) schematic reconstruction of neural pathways to the hypothalamic feeding area (HFA) in the goldfish brain, as inferred through degeneration techniques. It will be seen that the HFA receives both olfactory and gustatory information, and there is evidence that it also has a motor function (via the nuclei of the cranial nerves V and VII) to the muscles used in mouth movements.

Lastly, electrophysiological records have been taken of single or multiunit activity in the HFA of goldfish following stimulation of the vagal lobes and the olfactory tracts. Stimulation of either region led to increased firing rate of the HFA units, but when both stimuli were tested together the cells either fired in response to both stimuli or did not respond to either stimulus. This model is still tentative, and applies to teleosts which rely particularly on chemosensory mechanisms in feeding, but enough is now known to begin to explain the principal aspects of the hypothalamic control of teleost feeding (Demski, 1981). Although less is known of the control of feeding in elasmobranchs, the control system may be the same in both groups.

The cerebellum. In contrast to the cartilaginous fishes, there is a wide range in development of the cerebellum in other fishes, from the rudimentary structure in lampreys, to the enormous and elaborate structure in mormyrids (Figure 10.13). In lampreys, the cerebellum is simply a ridge of tissue anterior to the 4th ventricle, but in this, there are large Purkinje-like cells, and efferent fibres pass to the optic tectum, the tegmental and medullary motor nuclei, and to the oculomotor nucleus. Other efferents form descending cerebello-spinal tracts, or join the median longitudinal bundle of the reticular system (Niewenhuys, 1969). Even in lampreys, then, the cerebellum is built on the same sort of plan as in gnathostome fish and tetrapods.

Input to the cerebellar cortex in tetrapods is from two kinds of fibres only; climbing fibres, arising in the olive, that twine around the outspread dendrites of the Purkinje cells, and mossy fibres that synapse on the very numerous small granule cells. Climbing fibres have not yet been

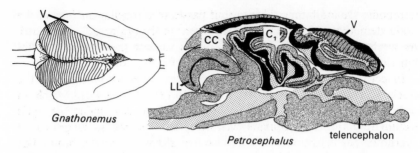

Gnathonemus

Petrocephalus telencephalon

Figure 10.13 The mormyrid cerebellum. Left: dorsal view of valvulae (V) completely covering brain in *Gnathonemus*. Right: near-sagittal section of brain of *Petrocephalus* showing corpus cerebelli (CC), lateral line lobe (LL), and valvula (V). C_1 is the first central lobe, whose neurones are seen in Figure 10.14. After Niewenhuys and Nicholson (1969).

unequivocally demonstrated in fishes. It seems unlikely that they can simply be resistant to the staining methods which demonstrate them in tetrapods, and most probably, there *are* 'climbing' fibres in fishes which do not climb, but simply terminate around the basal dendrites of the Purkinje cells. Granule cell axons rise to the molecular layer, where they bifurcate to form parallel fibres running orthogonally amongst the elaborate dendritic espaliers formed by the Purkinje and stellate cells.

Finally, in this complex architecture, there are Golgi cells, which synapse with the parallel fibres. The sole output is the Purkinje cell axons; except for the granule cells, all the neurones of the cerebellar cortex (at least in

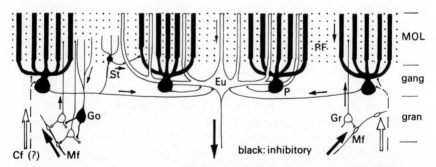

Figure 10.14 Cell types in lobe C_1 of the mormyrid cerebellum (*Gnathonemus*). Cells with inhibitory outputs, black. Note sole output is via axons of eurydendroid cells (Eu), on to which project Purkinje cell (P) axons. Climbing fibre input to the bases of the Purkinje cells is probable but not yet definitely established. Cf. climbing fibres; gang, ganglionic layer; Go, Golgi cells; Gr, granule cells; gran, granular layer; Mf, mossy fibres; MOL, molecular layer; St, stellate cells. Apart from the eurydendroid cells, the arrangement is similar to that of a normal cerebellum in a less specialized fish. After Niewenhuys, *et al.* (1974).

tetrapods) are inhibitory. This output passes to cerebellar nuclei and the brain stem, thence to motor output systems in the brain and spinal cord. As we have seen, experiments on dogfish suggest that the cerebellum functions to regulate the expression or gain of motor acts.

In certain advanced electrolocating fishes, such as mormyrids, the cerebellum is greatly enlarged (Figure 10.13); the valvula (which in ordinary teleosts forms a pouch projecting forwards under the optic tectum into the mesencephalic ventricle) has become so large that it overflows out of the ventricle and covers almost all of the brain. The corpus of the cerebellum is also enlarged and is differentiated into four distinct lobes. Figures 10.14 and 10.15 show something of the cell architecture and assumed mode of operation of one of these lobes in the mormyrid *Gnathonemus* (Niewenhuys *et al.*, 1974). As well as the usual elements found in the cerebellum of other forms, there are also so-called eurydendroid cells with wide dendritic trees which seem to form the sole output system from the lobe, since Purkinje cell axons terminate upon them (Figure 10.15).

Mormyrids possess several kinds of electroreceptors, and it seems that their input does not pass to the cerebellum directly (as does acoustico-lateralis input) but via somatotopic maps in the medulla and midbrain. Defined regions of the valvula receive indirect input from ampullary and small pore organs, other regions from large pore organs, and from the valvula there are many reciprocal connections with midbrain electro-sensory nuclei.

As we should expect, the cerebellum is in connection with the electric

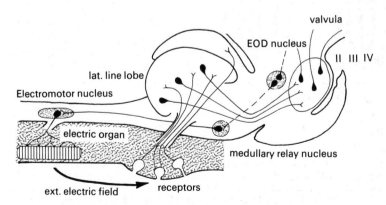

Figure 10.15 Relations between electric organ discharge and electroreceptors in a mormyrid. After Niewenhuys and Nicholson (1969).

organ discharge (EOD) centre, and its responses in mormyrids to electrical events depend upon the timing of the event in relation to the EOD centre command (Zipser and Bennett, 1976).

Mormyrids use their EOD not only for electrolocation and to identify objects of different conductivity in the field, but also for social purposes; not surprisingly the interaction between the input of the different receptor types and the EOD centre is complex and differs under different conditions. For example, in an unfamiliar situation, EOD rates increase, but when two mormyrids interact, the 'intruder' may inhibit its EOD, perhaps to conceal its identity (see Moller and Szabo, 1981). The unravelling of the central connections of the electroreceptor system in such highly evolved fishes is fascinating, and we only have space here to refer the reader to Bullock (1982) for an idea of the present understanding of electroreception in these and other fish. However, almost all fish groups have electroreceptive systems, and in elasmobranchs and other non-teleost groups, input enters by an anterior lateral line nerve and passes from the medulla to terminate in an octavolateral nucleus (Figure 10.16). Evidently, the receptor system is homologous in groups as different as lampreys and crossopterygians.

Brain centres for motor coordination. Fishes have extensive and elaborate muscles and so also have prominent centres in the brain for the control of movements. The simplest and most primitive are in the reticular formation of the midbrain and medulla. Here there are large neurones whose large rapidly-conducting axons descend to the spinal cord. The most complex and highly integrated system of motor control is in the cerebellum.

In lampreys, nine pairs of very large Müller cells and two pairs of 'Mauthner' cells have individual morphological and physiological features. There is also another distinct group of large reticular cells with axons in the lateral parts of the spinal cord. All these cells are synaptically connected through inputs from sensory systems that include the vestibular, the lateral line, the somatosensory, the visual and the olfactory (Rovainen, 1979).

The Müller cells, which are positioned in the lower half of the brain stem from the mid to hindbrain, send their large-diameter axons down the cord to synapse with interneurones and with motoneurones that innervate myotomal and dorsal-fin muscles. The Mauthner cells in the medulla are also involved in descending motor control. In both larval and adult lampreys, Rovainen found that responses from these cells lead to bilateral contraction of the myotomes. After continued stimulation the muscle

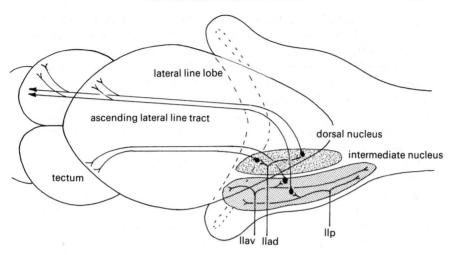

Figure 10.16 Electroreceptor input in dogfish. Anterior and posterior lateral line nerves (lla, llp) project to the intermediate nucleus whilst electrosensory input from ampullae of Lorenzini project to the dorsal nucleus in the lateral line lobe, and thence forwards to the midbrain. After Roberts *et al.* (1982).

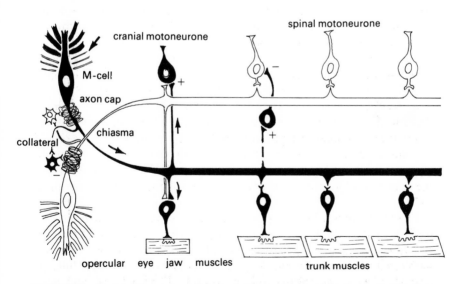

Figure 10.17 The Mauthner system. Schematic diagram showing Mauthner cells (M), their axons crossing at the Mauthner chiasma (C) before descending the cord and connecting with contralateral motoneurones. After Faber and Korn (1978).

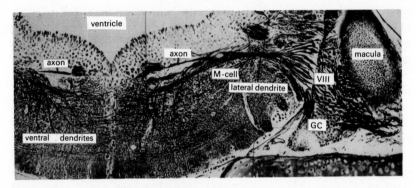

Figure 10.18 Mauthner cell of the lungfish *Protopterus*. Transverse reduced silver section of brain at level of nerve VIII. Note large fibre synaptic input to lateral dendrite from ganglion cells of nerve VIII (GC VIII); ventral dendrites; axons passing towards the chiasma; macula of inner ear to right, innervated from VIII.

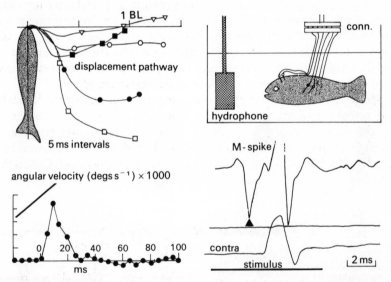

Figure 10.19 Responses of the Mauthner system. Upper left: displacement pathway of head of goldfish. Points indicate position of head at 5 ms intervals during 5 Mauthner responses. Lower left: angular velocity during Mauthner response of goldfish. Right: experimental arrangement, and record obtained from extra-cellular electrodes in hindbrain and in white zones of the myotomes of either side. Lower right: upper line, electrode near M-cell, arrow indicates M-spike (followed by a large movement artefact); middle line, ipsilateral muscle; lower line, contra-lateral muscle (activated). Solid bar below: duration of sound stimulus. Time-marker: 2 ms. After Zottoli (1977) and Eaton *et al.* (1977).

waves move forwards to induce behaviour similar to backward crawling. A Mauthner cell of the lamprey is excited directly by afferent fibres from the anterior and ipsilateral vestibular nerve, but stimulation of the contralateral nerve partner leads to excitation of the cell, not to its inhibition, as in teleosts. Rovainen concludes that either the lamprey cells are not strictly homologous with the M-cells of teleosts or, much more likely, they are homologous but have different functions.

In gnathostome fishes, the pair of Mauthner cells lies under the fourth ventricle in the medulla and opposite the vestibular nerves (VIII) (Figure 10.18). The cells are generally large (up to 100 μm in cross-section) and bear two principal dendrites, one dorso-lateral, the other ventro-rostral in position. Close to their origins, the two Mauthner axons, which are appreciably larger than others in the spinal cord and heavily myelinated, cross over and descend down the length of the spinal cord, then through collaterals form synapses with motoneurones and interneurones involved in the excitation of the myotomes (Figure 10.17). The anterior unmyelinated part of each M-axon is enveloped in a specialized axon cap. Each M-cell receives many afferent inputs from the vestibular (8th) nerves and there are inhibitory neurones associated with the axon cap. Dorsal and ventral acoustic nuclei, the principal nucleus of the trigeminal nerve and the tectobulbar and cerebellotegmental pathway have been tentatively identified as other sources of afferent connections to the M-cells.

The Mauthner system is present in lampreys (not hagfishes); in adult and juvenile Holocephali; in elasmobranchs (known only in embryo *Squalus*); the coelacanth *Latimeria*; lungfish; *Polypterus*; sturgeons; garfishes (*Lepisosteus*); and bowfin (*Amia*); and in most teleosts, but notably absent in eels, angler-fishes, pipe-fishes and sea-horses (Zottoli, 1981).

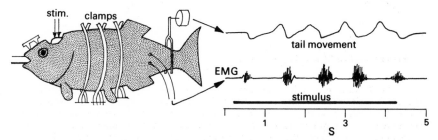

Figure 10.20 Experimental arrangement and records obtained by stimulation of the swimming centre in the teleost brain. Upper right: tail movement, lower right: electromyogram from caudal muscles. Solid bar: stimulus. After Kashin *et al.* (1974).

Electrophysiological tests show that the M-cells become active just before certain kinds of startle movements are made. For instance, in a fusiform fish, such as a goldfish, the startle response begins with either a C or S-type flexure of the body. The C-type start, usually called the 'tail-flip' is clearly associated with M-cell activity (Figure 10.19). The response, which has a latency as small as 5–8 ms, and starts with a C-bend, is followed by a strong counter-stroke of the tail so as to carry the fish away from its starting position. The crossing of the M-axons and their linkage to the myotomes is such that the head, the most vulnerable part of the body, is moved away from the side nearer to the disturbance, and so predators may be avoided (see Eaton and Bombardieri, 1978).

The swimming centre. In several teleost fishes, stimulation of a localized region in the caudal part of the midbrain tegmentum elicits coordinated swimming movements (Kashin *et al.*, 1974) suggesting that at least in these teleosts there is a mesencephalic locomotor region (MLR) just as in some mammals. Figure 10.20 shows the experimental arrangement, and some of the results obtained by Kashin and his colleagues, which perhaps imply that there are 'hard-wired' centres controlling stereotyped activities like feeding, swimming, and some aspects of reproductive behaviour such as nest-building. The Mauthner system provides a particularly simple example of such a hard-wired system. Another is given by the mesencephalic trigeminal nucleus system studied by Roberts and Witkovsky (1975) involved in the jaw closing reflex.

REFERENCES

Chapter 1

Avise, J. C and Kitto, G. G. (1973) Phosphoglucose isomerase gene duplication in the bony fishes: an evolutionary history. *Biochem. Gen.*, **8**, 113–132.

Bardack, D. and Zangerl, R. (1971) 'Lampreys in the fossil record', in *The Biology of Lampreys*, Vol. 1, Hardisty, M. W. and Potter, I. C., eds., Academic Press, London, pp. 67–84.

Compagno, L. J. V. (1977) Phyletic relationships of living sharks and rays. *Amer. Zool.*, **17**, 303–322.

Cohen, D. M. (1970) How many recent fishes are there? *Proc. Calif. Acad. Sci.*, **38**, 341–345.

Goodrich, E. S. (1909) 'Vertebrata craniata. Fasc. 1. Cyclostomes and fishes,' in *A Treatise on Zoology*, Lankester, E. Ray, ed., A. and C. Black, London.

Greenwood, P. H. (1973) 'Interrelationships of osteoglossomorphs', in *Interrelationships of Fishes*, Greenwood, P. H., Miles, R. S. and Patterson, C., eds., *Zool. J. Linn. Soc.*, **53**, suppl. 1, Academic Press, London, pp. 307–332.

Greenwood, P. H., Rosen, D. E., Weitzmann, S. H. and Myers, G. S. (1966) Phyletic studies of teleostean fishes with a provisional classification of living forms. *Bull. Amer. Mus. Nat. Hist.*, **131**, 339–456.

Hardisty, M. W. (1979) *Biology of the Cyclostomes*. Chapman and Hall, London.

Hetherington, T. E. and Bemis, W. E. (1979) Morphological evidence of an electroreceptive function of the rostral organ of *Latimeria chalumnae*. *Amer. Zool.*, **19**, 986.

Hinegardner, R. T. (1968) Evolution of cellular DNA content in teleost fishes. *Amer. Nat.*, **102**, 517–523.

Hinegardner, R. T. and Rosen, D. E. (1972) Cellular DNA content and the evolution of teleostean fishes. *Amer. Nat.*, **106**, 621–644.

Jarvik, E. (1980) *Basic Structure and Evolution of Vertebrates* (2 vols), Academic Press, London, pp. 1–575 and 1–337.

Lagler, K. F., Bardach, J. E., Miller, R. R., Passino, D. M. (1977) *Ichthyology* (2nd edn.). John Wiley and Sons, New York, pp. 1–506.

Lindsey, C. C. (1966) Body size of poikilotherm vertebrates at different latitudes. *Evolution*, **20**, 456–565

Locket, N. A. (1980) Some advances in coelacanth biology. *Proc. Roy. Soc. Lond. B*, **208**, 265–308.

Marshall, N. B. (1971) *Explorations in the Life of Fishes*. Harvard University Press, Cambridge, Mass.

Mayr, E. (1981) Biological classification: toward a synthesis of opposing methodologies. *Science*, **214**, 510–516.

Moss, M. L. (1977) Skeletal tissues in sharks. *Amer. Zool.*, **17**, 335–342.

Moss, S. A. (1977) Feeding mechanisms in sharks. *Amer. Zool.*, **17**, 355–364.

Moy-Thomas, J. A and Miles, R. S. (1971) *Paleozoic Fishes* (2nd edn.). Chapman and Hall, London.

Nelson, G. J. (1969) Gill arches and the phylogeny of fishes, with notes on the classification of vertebrates. *Bull. Amer. Mus. Nat. Hist.*, **141**, 475–552.

Nelson, J. S. (1976) *Fishes of the World.* John Wiley and Sons, New York.

Patterson, C. (1967) Are the teleosts a polyphyletic group? *Colloques Intl. C.N.R.S.* (Paris), **163**, 93–109.

Patterson, C. (1973) 'Interrelationships of holosteans', in *Interrelationships of Fishes*, Greenwood, P. H., Miles, R. S. and Patterson, C., eds., *Zool. J. Linn. Soc.*, **53**, suppl. 1, Academic Press, London, pp. 233–306.

Patterson, C. (1975) The braincase of pholidophorid and leptolepid fishes, with a review of the actinopterygian braincase. *Phil. Trans. Roy. Soc. Lond. B*, **190**, 275–579.

Patterson, C. (1977) 'The contribution of paleontology to teleostean phylogeny', in *Major Patterns in Vertebrate Evolution*, Held, M. K., Goody, P. C. and Hecht, B. M., eds. Plenum Press, New York, pp. 579–643.

Seigel, J. A. (1978) Revision of the dalatiid shark genus *Squaliolus*: anatomy, systematics, ecology. *Copeia*, 1978, 602–614.

Thomson, K. S., Gall, J. G. and Coggins, L. W. (1973) Nuclear DNA content of coelacanth erythrocytes. *Nature, Lond.*, **241**, 126.

Whiting, H. P. W. and Bone, Q. (1979) Ciliary cells in the epidermis of the larval Australian dipnoan, *Neoceratodus. Zool. J. Linn. Soc.*, **68**, 125–137.

Young, J. Z. (1981) *The Life of Vertebrates* (3rd edn.) Clarendon Press, Oxford.

Chapter 2

Alexander, R. McN. (1965) Structure and function in the catfish. *J. Zool.* (Lond.), **148**, 88–152.

Bone, Q. and Roberts, B. L. (1969) The density of elasmobranchs. *J. mar. biol. Ass. UK*, **49**, 913–937.

Briggs, J. C. (1974) *Marine Zoogeography.* McGraw-Hill, New York.

Brittan, M. R. (1961) Adaptive radiation in Asiatic cyprinid fishes, and their comparison with forms from other areas. *Proc. 9th Pacific Sci. Congress*, **10** (Fisheries), 18–31.

Carey, F. G., Teal, J. M., Kanwisher, J. W., Lawson, K. D. and Beckett, J. S. (1971) Warm-bodied fish. *Amer. Zool.*, **11**, 137–145.

Cohen, D. M. (1970) How many recent fishes are there? *Proc. Calif. Acad. Sci.*, **38**, 341–345.

De Vries, A. L. (1971) 'Freezing resistance in fishes', in *Fish Physiology*, Vol. 6, Hoar, W. S. and Randall, D. J., eds., Academic Press, New York and London, pp. 157–190.

De Vries, A. L. (1974) 'Survival at freezing temperatures', in *Biochemical and Biophysical Perspectives in Marine Biology*, Vol. 1, Malins, D. C. and Sargent, J. R., eds., Academic Press, London, pp. 289–333.

Dijkgraaf, S. (1960) Hearing in bony fishes. *Proc. Roy. Soc. Lond. B*, **152**, 51–54.

Dobbs, G. H., Lin, Y. and De Vries, A. L. (1974) Aglomerularism in Antarctic fish. *Science*, **185**, 793–794.

Fryer, G. and Iles, T. D. (1972) *The Cichlid Fishes of the Great Lakes of Africa.* Oliver and Boyd, Edinburgh.

Hochachka, P. W. (1980) *Living Without Oxygen.* Harvard University Press, Cambridge, Mass.

Horn, M. H. (1972) The amount of space available for marine and freshwater fishes. *Fish. Bull.*, **70**, 1295–1297.

Lowe-McConnell, R. H. (1975) *Fish Communities in Tropical Freshwaters.* Longman, London.

Marshall, N. B. (1965) *The Life of Fishes.* Weidenfeld and Nicolson, London.

Marshall, N. B. (1971) *Explorations in the Life of Fishes*. Harvard University Press, Cambridge, Mass.

Marshall, N. B. (1979) *Developments in Deep-Sea Biology*. Blandford Press, Poole.

Miller, R. R. (1950) Speciation in fishes of the genera *Cyprinodon* and *Empertrichthys* inhabiting the Death Valley region. *Evolution*, **4**, 155–163.

Pfeiffer, W. (1962) The fright reaction of fish. *Biol. Revs.*, **37**, 495–511.

Pfeiffer, W. (1977) The distribution of fright reaction and alarm substances in fishes. *Copeia*, 1977, 653–665.

Pillay, T. V. R. (1967) 'Estuarine fishes of West Africa', and 'Estuarine fishes of the Indian Ocean coastal zone', in *Estuaries*, Lauff, G. H., ed., publ. no. 83, Amer. Ass. Adv. Sci., Washington D.C., pp. 639–649, 647–657.

Regan, C. T. (1936) 'Pisces', in *Natural History*, Regan, C. T., ed. Ward Lock and Co., London.

Sale, P. F. (1980) The ecology of fishes on coral reefs. *Oceanography Mar. Biol. Ann. Rev.*, **18**, 367–421.

Sulak, K. J. (1977) The systematics and biology of *Bathypterois* (Pisces, Chloropthalmidae) with a revised classification of benthic myctophiform fishes. *Galathea report*, **14**, 49–108.

Weitzmann, S. H. (1962) The osteology and relationships of the South American characid fishes of the subfamily Gasteropelecinae. *Stanford Ichthyol. Bull.*, **4**, 211–263.

Wishner, K. F. (1980) The bio-mass of the deep-sea benthopelagic plankton. *Deep Sea Res.*, **27A**, 203–216.

Yoakim, E. G. and Grizzle, J. M. (1982) Ultrastructure of alarm substance cells in the epidermis of the channel catfish, *Ictalurus punctatus* (Rafinesque). *J. Fish. Biol.*, **20**, 213–221.

Chapter 3

Alexander, R. McN. (1969) The orientation of muscle fibres in the myomeres of fishes. *J. mar. biol. Ass. U.K.*, **49**, 263–290.

Bone, Q. (1972) Buoyancy and hydrodynamic functions of integument in the castor oil fish, *Ruvettus pretiosus* (Pisces: Gempylidae). *Copeia*, 1972, 78–87.

Bone, Q. (1978) 'Locomotor muscle', in *Fish Physiology*, Vol. 7, Hoar, W. S. and Randall, D. J., eds., Academic Press, New York, pp. 361–424.

Bone, Q., Kiceniuk, J. and Jones, D. R. (1978) On the role of the different fibre types in fish myotomes at intermediate swimming speeds. *Fish. Bull. U.S.*, **76**, 691–699.

Carey, F. G. and Teal, J. M. (1966) Heat conservation in tuna fish muscle. *Proc. Natl. Acad. Sci. USA*, **56**, 1464–1469.

Carey, F. G. and Teal, J. M. (1969) Mako and porbeagle: warm-bodied sharks. *Comp. Biochem. Physiol.*, **28**, 199–204.

Carey, F. G., Teal, J. M., Kanwisher, J., Lawson, K. D. and Beckett, J. S. (1971) Warm-bodied fish. *Amer. Zool.*, **11**, 137–145.

Castellani, M. A. and Somero, G. N. (1981) Buffering capacities of vertebrate muscle: correlations with potentials for anaerobic function. *J. comp. Physiol.*, **143**, 191–198.

Gray, J. (1933) Studies in animal locomotion. I. The movement of fish with special reference to the eel. *J. exp. Biol.*, **10**, 88–104.

Lighthill, J. (1971) Large-amplitude elongated body theory of fish locomotion. *Proc. Roy. Soc. Lond. B*, **179**, 125–138.

Nursall, J. R. (1956) The lateral musculature and the swimming of fish. *Proc. zool. soc. Lond.*, **126**, 127–143.

Smit, H., Amelink-Koutstall, J. M., Vijverberg, J. and von Vaupel-Klein, J. C. (1971) Oxygen consumption and efficiency of swimming goldfish. *Comp. Biochem. Physiol.*, **39A**, 1–28.

Walters, V. (1963) The trachypterid integument and an hypothesis on its hydrodynamic function. *Copeia*, 1963, 260–270.

Wardle, C. S. (1975) Limits of fish swimming speed. *Nature*, **255**, 725–727.

Wardle, C. S. and Videler, J. J. (1979) How do fish break the speed limit? *Nature*, **284**, 445–447.

Webb, P. W. (1971) The swimming energetics of trout. I. Thrust and power output at cruising speeds. *J. exp. Biol.*, **55**, 489–520.

Webb, P. W. (1978) 'Hydrodynamics: nonscombroid fish', in *Fish Physiology*, Vol. 7, Hoar, W. S. and Randall, D. J., eds., Academic Press, New York, pp. 189–237.

Chapter 4

Alexander, R. McN. (1966) Physical aspects of swimbladder function. *Biol. Revs.*, **41**, 141–176.

Alexander, R. McN. (1972) The energetics of vertical migration by fishes. *Symp. Soc. exp. Biol.*, **26**, 273–294.

Baldridge, H. D. (1970) Sinking factors and average densities of Florida sharks as functions of liver buoyancy. *Copeia*, 1970, 744–754.

Berg, T. and Steen, J. B. (1968) The mechanism of oxygen concentration in the swimbladder of the eel. *J. Physiol., Lond.*, **195**, 631–638.

Blaxter, J. H. S. and Tytler, P. (1978) Physiology and function of the swimbladder. *Adv. comp. Physiol. Biochem.*, **7**, 311–367.

Bone, Q. (1973) A note on the buoyancy of some lantern fishes (Myctophidae). *J. mar. biol. Ass. U.K.*, **53**, 619–633.

Bone, Q. and Roberts, B. L. (1969) The density of elasmobranchs. *J. mar. biol. Ass. U.K.*, **49**, 913–937.

Brawn, V. M. (1962) Physical properties and hydrostatic function of the swimbladder of herring (*Clupea harengus* L.). *J. Fish. Res. Bd. Can.*, **19**, 635–656.

Butler, J. L. and Pearcey, W. G. (1972) Swimbladder morphology and specific gravity of myctophids off Oregon. *J. Fish. Res. Bd. Can.*, **29**, 1145–1150.

Corner, E. D. S., Denton, E. J. and Forster, G. R. (1969) On the buoyancy of some deep-sea sharks. *Proc. Roy. Soc. Lond. B.*, **171**, 415–429.

Denton, E. J. (1961) The buoyancy of fish and cephalopods. *Prog. Biophys. biophys. Chem.*, **11**, 178–234.

Denton, E. J. and Marshall, N. B. (1958) The buoyancy of bathypelagic fishes without a gas-filled swimbladder. *J. mar. biol. Ass. U.K.*, **37**, 753–767.

Denton, E. J., Liddicoat, J. D. and Taylor, D. W. (1972) The permeability to gases of the swimbladder of the conger eel (*Conger conger*). *J. mar. biol. Ass. U.K.*, **52**, 727–746.

Eastman, J. J. and deVries, A. L. (1981) Buoyancy adaptations in a swimbladder-less Antarctic fish. *J. morphol.*, **167**, 91–102.

Fänge, R. (1966) Physiology of the swimbladder. *Physiol. Revs.*, **46**, 299–322.

Kanwisher, J. and Ebeling, A. (1957) Composition of the swimbladder gas in bathypelagic fishes. *Deep-sea Res.*, **4**, 211–217.

Krogh, A. (1919) The rate of diffusion of gases through animal tissues, with some remarks on the coefficient of invasion. *J. Physiol., Lond.*, **52**, 391–408.

Malins, D. C. and Barone, A. (1970) Glyceryl ether metabolism: regulation of buoyancy in dogfish, *Squalus acanthias*. *Science*, **167**, 79–80.

Marshall, N. B. (1960) Swimbladder structure of deep-sea fishes in relation to their systematics and biology. *Discovery Reports*, **31**, 1–122.

Marshall, N. B. (1972) Swimbladder organization and depth ranges of deep sea teleosts. *Symp. Soc. exp. Biol.*, **26**, 261–272.

Nevenzel, J. C., Rodegker, W., Robinson, J. S. and Kayama, M. (1969) The lipids of some lantern fishes. *Comp. Biochem. Physiol.*, **31**, 25–36.

Chapter 5

Bone, Q. (1978) 'Myotomal muscle fibre types in *Scomber* and *Katsuwonus*', in *The Physiological Ecology of Tunas*, Sharp, G. D. and Dizon, A. E., eds., Academic Press, New York, pp. 183–205.

Brown, C. E. and Muir, B. S. (1970) Analysis of ram ventilation in fish gills with application to skipjack tuna (*Katsuwonus pelamis*). *J. Fish. Res. Bd. Can.*, **27**, 1637–52.

Dunel, S. and Laurent, P. (1980) 'Functional organisation of the gill vasculature in different classes of fish', in *Epithelial Transport in the Lower Vertebrates*, Lahlou, B., ed., Cambridge University Press, pp. 37–58.

Farmer, G. J. and Beamish, F. W. H. (1969) Oxygen consumption of *Tilapia nilotica* in relation to swimming speed and salinity. *J. Fish. Res. Bd. Can.*, **26**, 2807–2821

Greene, C. W. (1902) Contributions to the physiology of the Californian hagfish *Polistotrema stouti*. II. The absence of regulative nerves for the systemic heart. *Amer. J. Physiol.*, **6**, 318–324.

Haswell, G. M. and Randall, D. J. (1978) The pattern of carbon dioxide excretion in the rainbow trout *Salmo gairdneri*. *J. exp. Biol.*, **72**, 17–24.

Hughes, G. M. (1970) A comparative approach to fish respiration. *Experientia*, **26**, 113–122.

Hughes, G. M. and Ballantijn, C. M. (1965) The muscular basis of the respiratory pumps in the dogfish (*Scyliorhinus canicula*). *J. exp. Biol.*, **43**, 363–383.

Hughes, G. M. and Morgan, M. (1973) The structure of fish gills in relation to their respiratory function. *Biol. Revs.*, **48**, 419–475.

Johansen, K. (1970) 'Air breathing in fishes', in *Fish Physiology*, Vol. 4, Hoar, W. S. and Randall, D. J., eds., Academic Press, New York, pp. 361–411.

Johansen, K. and Strahan, R. (1963) 'The respiratory system of *Myxine glutinosa* L.', in *The Biology of Myxine*, Brodal, A. and Fänge, R., eds., Universitetsforlaget, Oslo, pp. 352–371.

Johansen, K., Lenfant, C. and Hanson, D. (1968) Cardiovascular dynamics in the lungfishes. *Z. vergleich. Physiol.*, **59**, 157–186.

Johansen, K., Lenfant, C., Schmidt-Nielsen, K. and Petersen, J. (1968) Gas exchange and control of breathing in the electric eel, *Electrophorus electricus*. *Z. vergl. Physiol.*, **61**, 137–163.

Magnuson, J. J. (1978) 'Locomotion by scombroid fishes—hydromechanics, morphology and behaviour', in *Fish Physiology*, Vol. 7, Hoar, W. S. and Randall, D. J., eds., Academic Press, New York, pp. 239–313.

Powers, D. A. (1980) Molecular ecology of teleost fish hemoglobins: strategies for adapting to changing environments. *Amer. Zool.*, **20**, 139–162.

Roberts, J. L. (1975) Active branchial and ram gill ventilation in fishes. *Biol. bull. Woods Hole*, **148**, 85–105.

Steen, J. B. and Kruysse, A. (1964) The respiratory function of teleostean gills. *Comp. Biochem. Physiol.*, **12**, 127–142.

Stevens, E. D., Bennion, G. R., Randall, D. J. and Shelton, G. (1972) Factors affecting arterial pressures and blood flow from the heart in the intact unrestrained lingcod *Ophiodon elongatus*. *Comp. Biochem. Physiol.*, **43A**, 681–695.

Sudak, F. N. (1965) Intrapericardial and intracardiac pressures and the events of the cardiac cycle in *Mustelus canis* (Mitchell). *Comp. Biochem. Physiol.*, **15**, 199–215.

Vogel, W., Vogel, V. and Pfautsch, M. (1976) Arterio-venous anastamoses in rainbow trout gill filaments. *Cell Tiss. Res.*, **167**, 373–387.

Vogel, W. and Kock, K. H. (1981) Morphology of gill vessels in icefish. *Arch. Fisch. Wiss.*, **31**, 139–150.

Chapter 6

Altringham, J. D., Yancey, P. H. and Johnston, I. A. (1982) The effect of osmoregulatory solutes on tension generation by dogfish skinned muscle fibres. *J. exp. Biol.*, **96**, 443–445.

Burger, J. W. and Hess, W. N. (1960) Function of the rectal gland in the spiny dogfish, *Science*, **131**, 670–671.

Feldmeth, C. R. and Waggoner, J. P. (1972) Field measurements of tolerance to extreme hypersalinity in the California killifish, *Fundulus parvipinnis*. *Copeia*, 1972, 592–594.

Foskett, J. B. and Scheffey, C. (1982) The chloride cell: definitive identification as the salt-secretory cell in teleosts. *Science*, **215**, 164–166.

Griffith, R. W. and Thomson, R. S. (1973) *Latimeria chalumnae*: reproduction and conservation. *Nature*, **242**, 617–618.

Griffith, R. W., Pang, P. K. T., Srivastava, A. K. and Pickford, G. E. (1973) Serum composition of freshwater stingrays (Potamotrygonidae) adapted to fresh and dilute sea water. *Biol. Bull. Woods Hole*, **144**, 304–320.

Griffith, R. W. and Pang, P. K. T. (1979) 'Mechanisms of osmoregulation in the coelacanth: evolutionary implications', in *The Biology and Physiology of the Living Coelacanth*, McCosker, J. E. and Lagios, M. D., eds., *Occ. papers Calif. Acad. Sci.*, no. 134, pp. 79–93.

Hardisty, M. W. (1979) *Biology of the Cyclostomes*. Chapman and Hall, London.

Haywood, G. P. (1975) A preliminary investigation into the roles played by the rectal gland and kidneys in the osmoregulation of the striped dogfish *Poroderma africanum*. *J. exp. zool.*, **193**, 167–175.

Hickman, C. P. (1965) Studies on renal function in freshwater teleost fish. *Trans. Roy. Soc. Can.*, sect. 3, **3**, 213–236.

Hickman, C. P. and Trump, B. F. (1969) 'The kidney', in *Fish Physiology*, Vol. 1, Hoar, W. S. and Randall, D. J., eds., Academic Press, New York, pp. 91–240.

Kirsch, R., Meens, R. and Meister, M. F. (1981) Osmoregulation chez les teleostéens marins: rôle des branchies et du tube digestif. *Bull. Soc. zool. France*, **106**, 31–36.

Krogh, A. (1939) *Osmotic Regulation in Aquatic Animals*. Cambridge University Press.

Lagios, M. (1979) 'The coelacanth and the chondrichthyes as sister groups: a review of shared apomorph characters and a cladistic analysis and reinterpretation', in *The Biology and Physiology of the Living Coelacanth*, McCosker, J. E. and Lagios, M. D., eds., *Occ. Papers Calif. Acad. Sci.*, no. 134, pp. 25–44. See also the rebuttal of Lagios' thesis by Compagno, L. J. V. and Lagios' reply to Compagno, *ibid*, pp. 45–52 and 53–55.

Lutz, P. L. (1975) Adaptive and evolutionary aspects of the ionic content of fishes. *Copeia*, 1975, 369–373.

McInerney, J. E. (1974) Renal sodium reabsorption in the hagfish *Eptatretus stouti*. *Comp. biochem. Physiol.*, **49A**, 273–280.

Maetz, J. (1969) Seawater teleosts: evidence for a sodium-potassium exchange in the branchial sodium-excretory pump. *Science*, **166**, 613–615.

Maetz, J. (1974) 'Aspects of adaptation to hypo-osmotic and hyper-osmotic environments', in *Biochem. biophys. Perspect. Mar. biol.*, **1**, Malins, D. C. and Sargent, J. R., eds., pp. 1–167.

Munz, R. W. and McFarland, W. N. (1964) Regulatory function of a primitive vertebrate kidney. *Comp. biochem. Physiol.*, **13**, 381–400.

Pang, P. K. T., Griffith, R. W. and Atz, J. W. (1977) Osmoregulation in elasmobranchs. *Amer. Zool.*, **17**, 365–377.

Payan, P. and Maetz, J. (1973) Branchial sodium transport mechanisms in *Scyliorhinus canicula*: evidence for Na^+/NH_4^+ and Na^+/H^+ exchange and for a role of carbonic anhydrase. *J. exp. Biol.*, **58**, 487–502.

Pickering, A. D. and Morris, R. (1970) Osmoregulation of *Lampetra fluviatilis* and *Petromyzon marinus* (Cyclostomata) in hypertonic solutions. *J. exp. Biol.*, **53**, 231–243.

Potts, W. T. W. (1976) 'Ion transport and osmoregulation in marine fish', in *Perspectives in Experimental Biology, 1, Zoology*. Davies, P. S., ed., Pergamon Press, Oxford, pp. 65–75.

Ritchie, A. (1968) New evidence on *Jamoytius kerwoodi* White., an important ostracoderm from the Silurian of Lanarkshire, Scotland. *Paleontology*, **11**, 21–38.

Sardet, C., Pisam, M. and Maetz, J. (1980) 'Structure and function of gill epithelium of euryhaline teleost fish', in *Epithelial Transport in the Lower Vertebrates*, Lahlou, B., ed., Cambridge University Press, pp. 59–68.

Sargent, J. R., Pirie, B. J. S., Thomson, A. J. and George, S. G. (1978) 'Structure and function of chloride cells in the gills of *Anguilla anguilla*', in *Physiology and Behaviour of Marine Organisms*, McLusky, D. S. and Berry, A. J., eds., Pergamon Press, Oxford, pp. 123–132.

Smith, H. W. (1929) The excretion of ammonia and urea by the gills of fish. *J. biol. Chem.*, **81**, 727–742.

Smith, H. W. (1930) The absorption and secretion of water and salts by marine teleosts. *Amer. J. Physiol.*, **93**, 480–505.

Smith, H. W. (1931) The absorption and secretion of water and salts by the elasmobranch fishes. I. Freshwater elasmobranchs. *Amer. J. Physiol.*, **98**, 279–295.

Smith, H. W. (1932) Water regulation and its evolution in the fishes. *Quart. Rev. Biol.*, **7**, 1–26.

Stolte, H., Galaske, R. G., Eisenbach, G. M., Lechene, C., Schmidt-Nielsen, B. and Boylan, J. W. (1977) Renal tubule ion transport and collecting duct function in the elasmobranch little skate *Raja erinacea*. *J. exp. Zool.*, **199**, 403–410.

Thorson, T. B., Cowan, C. M. and Watson, D. E. (1967) *Potamotrygon* spp.: elasmobranchs with low urea content. *Science*, **158**, 375–377.

Yancey, P. H. and Somero, G. N. (1978) Urea-requiring lactate dehydrogenases of marine elasmobranch fishes. *J. comp. Physiol.*, **125**, 135–141.

Yancey, P. H. and Somero, G. N. (1979) Counteraction of urea destabilization of protein structure by methylamine osmoregulatory compounds of elasmobranch fishes. *Biochem. J.*, **183**, 317–323.

Chapter 7

Alexander, R. McN. (1970) Mechanics of the feeding action of various teleost fishes. *J. Zool., Lond.*, **162**, 145–156.

Bigelow, H. B. and Schroeder, W. C. (1948) *Fishes of the Western North Atlantic. I. Lancelets, Cyclostomes, Sharks.* (Publ. Sears Foundation Mar. Res., Memoir 1,) New Haven.

Fryer, G. and Iles, T. D. (1972) *The Cichlid Fishes of the Great Lakes of Africa.* Oliver and Boyd, Edinburgh.

Hickling, C. F. (1961) *Tropical Inland Fisheries.* Longman, London.

Hobson, E. S. (1974) Feeding relationships of teleostean fishes on coral reefs. *Fish Bull. U.S.*, **72**, 915–1031.

Hobson, C. F. and Chess, J. R. (1976) Trophic interactions among fishes and zooplankters near shore at Santa Catalina Island, California. *Fish. Bull. U.S.*, **74**, 567–598.

Hunter, J. R. (1972) Swimming and feeding behaviour of larval anchovy (*Engraulis mordax*). *Fish. Bull. U.S.*, **70**, 821–838.

Hyatt, K. D. (1979) 'Feeding strategy', in *Fish Physiology*, Vol. 3, Hoar, W. S., Randall, D. J. and Brett, J. R., eds., Academic Press, New York, pp. 71–119.

Inger, R. F. and Kong, C. P. (1962) The freshwater fishes of North Borneo. *Fieldiana: Zool.*, **45**, 1–268.

Keast, A. (1970) 'Food specialization and bioenergetic interrelations in the fish faunas of some small Ontario waterways', in *Marine Food Chains*, Steele, J. H., ed., Oliver and Boyd, Edinburgh, pp. 377–411.

Lauder, G. V. (1979) Feeding mechanisms in primitive teleosts and in the halecomorph fish *Amia calva*. *J. Zool., Lond.*, **187**, 543–578.

Lauder, G. V. (1980*b*) The suction feeding mechanism in sunfishes (*Lepomis*): an experimental analysis. *J. exp. Biol.*, **88**, 49–72.

Lauder, G. V. (1980) Evolution of the feeding mechanism in primitive actinopterygian fishes: a functional anatomical analysis of *Polypterus*, *Lepisosteus* and *Amia*. *J. Morphol.*, **163**, 283–317.

Lauder, G. V. and Liem, K. F. (1980) 'The feeding mechanism and cephalic myology of *Salvelinus fontinalis*: form, function and evolutionary significance', in *Charrs: salmonid fishes of the genus Salvelinus*, Balon, E. K., ed., W. Junk, The Hague, pp. 365–390.

Lauder, G. V. and Liem, K. F. (1981) Prey capture by *Luciocephalus pulcher*: implications for models of jaw protrusion in fishes. *Env. Biol. Fish*, **6**, 257–268.

Liem, K. F. (1980*a*) Adaptive significance of intra- and interspecific differences in the feeding repertoires of cichlid fishes. *Amer. Zool.*, **20**, 295–314.

244 BIOLOGY OF FISHES

Liem, K. F. (1980*b*) 'Acquisition of energy by teleosts: adaptive mechanisms and evolutionary patterns', in *Environmental Physiology of Fishes*, Ali, M. A., ed., Plenum, New York, pp. 299–334.

Liem, K. F. and Greenwood, P. H. (1981) A functional approach to the phylogeny of the pharyngognath teleosts. *Amer. Zool.*, **21**, 83–101.

Losey, G. S. (1972) The ecological importance of cleaning symbiosis. *Copeia*, 1972, 820–833.

Mallatt, J. (1981) The suspension feeding mechanism of the larval lamprey, *Petromyzon marinus*. *J. Zool., Lond.*, **194**, 103–142.

Moss, S. A. (1977) Feeding mechanisms in sharks. *Amer. Zool.*, **17**, 355–364.

Nelson, G. J. (1967) Epibranchial organs in lower teleostean fishes. *J. Zool., Lond.*, **153**, 71–89.

Nikolsky, G. V. (1963) *The Ecology of Fishes* (translated from the Russian by L. Birkett). Academic Press, London.

Nursall, J. R. (1981) The activity budget and use of territory by a tropical blenniid fish. *Zool. J. Linn. Soc.*, **72**, 69–92.

Parker, H. W. and Boeseman, M. (1954) The basking shark *Cetorhinus maximus* in winter. *Proc. zool. Soc. Lond.*, **124**, 185–194.

Pietsch, T. W. (1974) Osteology and relationships of ceratioid angler fishes of the family Oneirodidae, with a review of the genus *Oneirodes* Lutken. *Bull. Nat. Hist. Mus. Los Angeles County, Science*, **18**, 1–112.

Pietsch, T. W. and Grobecker, D. B. (1978) The compleat angler: aggressive mimicry in an antennariid angler fish. *Science*, **201**, 369–370.

Randall, J. E. (1967) 'Food habits of reef fishes of the West Indies', in *Proc. Int. conference on Tropical Oceanography (1965)*, University of Miami, Institute of Marine Sciences, pp. 665–847.

Schaeffer, B. and Rosen, D. E. (1961) Major adaptive levels in the evolution of the actinopterygian feeding mechanism. *Amer. Zool.*, **1**, 187–204.

Shelbourne, J. E. (1962) A predator–prey size relationship for plaice larvae feeding on *Oikopleura*. *J. mar. biol. Ass. U.K.*, **42**, 243–252.

Webb, P. W. and Skadsen, J. M. (1980) Strike tactics of *Esox*. *Can. J. Zool.*, **58**, 1462–1469.

Weihs, D. and Moser, H. G. (1981) Stalked eyes as an adaptation towards more efficient foraging in marine fish larvae. *Bull. mar. Sci.*, **31**, 31–36.

Wickler, W. (1968) *Mimicry in Plants and Animals*. Weidenfeld and Nicolson, London.

Young, J. Z. (1981) *The Life of Vertebrates* (3rd edn.) Clarendon Press, Oxford.

Zaret, T. M. (1980) *Predation and Freshwater Communities*. Yale University Press, New Haven.

Chapter 8

Badcock, J. and Merrett, N. R. (1976) Midwater fishes in the eastern North Atlantic. I. Vertical distribution and associated biology in 30°N, 23°W, with developmental notes on certain myctophids. *Progr. Oceanogr.*, **7**, 3–58.

Breder, C. M. (1962) On the significance of transparency in osteichthyid fish eggs and larvae. *Copeia*, 1962, 561–567.

Bruce, R. W. (1979) *A Study of the Scaridae of Aldabra Atoll*. Ph.D. thesis, University of Glasgow.

Clarke, A. (1980) A reappraisal of the concept of metabolic cold adaptation in polar marine invertebrates. *Biol. J. Linn. Soc.*, **14**, 77–92.

Cody, M. (1966) A general theory of clutch size. *Evolution*, **20**, 174–184.

Cushing, D. H. (1975) *Marine Ecology and Fisheries*. Cambridge University Press.

Denton, E. J. (1963) Buoyancy mechanisms of sea creatures. *Endeavour*, **22**, 3–8.

Everson, I. (1977) *The Living Resources of the Southern Ocean*. Southern Ocean Fish. Survey Programme GLO/SO/77/1, FAO, Rome, pp. 156.

Fischer, E. A. (1981) Sexual allocation in a simultaneously hermaphrodite coral reef fish. *Amer. Nat.*, **117**, 64–82.

Francis, M. P. (1981) Von Bertalanffy growth rates in species of *Mustelus* (Elasmobranchii: Triakidae). *Copeia*, 1981, 189–192.

Ghiselin, M. T. (1969) The evolution of hermaphroditism among animals. *Quart. Rev. Biol.*, **44**, 189–208.

Gould, S. J. (1977) *Ontogeny and Phylogeny*. Harvard University Press, Cambridge, Mass.

Harden-Jones, F. R. (1968) *Fish Migration*. Edward Arnold, London.

Hardisty, M. W. (1979) *Biology of the Cyclostomes*. Chapman and Hall, London.

Hardisty, M. W. and Potter, I. C. (1971) 'The behaviour, ecology and growth of larval lampreys', in *The Biology of Lampreys*, Vol. 1, Hardisty, M. W. and Potter, I. C., eds., Academic Press, London, pp. 85–125.

Harrington, R. W. (1971) How ecological and genetic factors interact to determine when self-fertilising hermaphrodites of *Rivulus marmoratus* change into functional secondary males, with a reappraisal of the modes of intersexuality among fishes. *Copeia*, 1971, 389–432.

Hoar, W. S. (1969) 'Reproduction', in *Fish Physiology*, Vol. 3, Edar, W. S. and Randall, D. J., eds., Academic Press, New York, pp. 1–72.

Holden, M. J. (1974) 'Problems in the rational exploitation of elasmobranch populations and some suggested solutions', in *Sea Fisheries Research*, Harden-Jones, F. R., ed., Elek Science, London, pp. 117–137.

Holmgren, N. (1947) On two embryos of *Myxine glutinosa*. *Acta Zool. (Stockh.)*, **27**, 1–90.

Hureau, J-C. and Ozouf, C. (1977) Détermination de l'âge et croissance du coelacanth *Latimeria chalumnae* Smith 1939 (Poisson, Crossopterygien, Coelacanthidé). *Cybium*, **2**, 129–137.

Kawaguchi, K. and Marumo, R. (1967) Biology of *Gonostoma gracile* (Gonostomatidae) Part I. *Information Bull. on Planktonol. in Japan* (issue in commem. of Dr. Y. Matsue), pp. 53–67.

Krumholz, L. A. (1948) Reproduction in the western mosquito fish *Gambusia affinis affinis* (Baird and Girard), and its use in mosquito control. *Ecol. Monogr.*, **18**, 1–43.

Lowe-McConnell, R. H. (1975) *Fish Communities in Tropical Freshwaters*. Longman, London and New York.

MacArthur, R. H. and Wilson, E. O. (1967) *Theory of Island Biogeography*. Princeton University Press, Princeton.

Marshall, N. B. (1953) Egg size in Arctic, Antarctic and deep-sea fishes. *Evolution*, **7**, 328–41

Marshall, N. B. (1965) *The Life of Fishes*. Weidenfeld and Nicolson, London.

Marshall, N. B. (1979) *Developments in Deep-sea Biology*. Blandford Press, Poole.

Miller, P. J. (1979) Adaptiveness and implications of small size in teleosts. *Symp. zool. Soc. Lond.*, **44**, 263–306.

Parry, G. D. (1981) The meanings of r- and K-selection. *Oecologia (Berl.)*, **48**, 260–264.

Por, F. D. (1978) *Lessepsian Migration*. Springer-Verlag, Berlin.

Regan, C. T. (1916) Larval and post-larval fishes. *Brit. Antarct. Terra Nova Exped. 1910* (*Zoology*, 1).

Rosen, D. E. (1973) 'Suborder Cyprinodontoidei, superfamily Cyprinodontoidea, families Cyprinodontidae, Poeciliidae, Anablepidae', in *Fishes of the Western North Atlantic*, no. 1, pt. 6. Sears Foundation Mar. Res., Yale University, pp. 229–262.

Russell, F. S. (1976) *The Eggs and Planktonic Stages of British Marine Fishes*. Academic Press, London.

Sale, P. F. (1980) The ecology of fishes on coral reefs. *Oceanogr. mar. biol. Ann. Rev.*, **18**, 367–422.

Schultz, R. J. (1977) 'Evolution and ecology in unisexual fishes', in *Evolutionary Biology*, Vol. 10, Plenum Press, New York, pp. 277–331.

Simpson, B. R. C. (1979) The phenology of annual killifishes. *Symp. zool. Soc. Lond.*, **44**, 243–261.

Stearns, S. C. (1976) Life history tactics: a review of the ideas. *Quart Rev. Biol.*, **51**, 3–65.

Sulak, K. J. (1977) The systematics and biology of *Bathypterois* (Pisces: Chloropthalmidae) with a revised classification of benthic myctophiform fishes. *Galathea report*, **14**, 49–108.

Walvig, F. (1963) 'Gonads and formation of sexual cells', in *The Biology of Myxine*, Brodal, A. and Fänge, R., eds., Universitetsforlaget, Oslo, Norway, pp. 530–580.

Whiting, H. P. W. (1972) 'The organisation of larval lampreys', in *Studies in Vertebrate Evolution*, Joysey, K. and Kemp, T. S., eds., Oliver and Boyd, Edinburgh, pp. 1–20.

Wilbur, H. M. (1980) Complex life-cycles. *Ann. Rev. Ecol. Syst.*, **11**, 67–93.

Wourms, J. P. (1977) Reproduction and development in chondrichthyan fishes. *Amer. Zool.*, **17**, 379–410.

Wyatt, T. (1980) The growth season in the sea. *J. Plankton Res.*, **2**, 81–96.

Chapter 9

Baguet, F., Christophe, B. and Marechal, G. (1980) Luminescence of *Argyropelecus* photophores electrically stimulated. *Comp. Biochem. Physiol.*, **67A**, 375–381

Banner, A. (1967) 'Evidence of sensitivity to acoustic displacements in the lemon shark, *Negaprion brevirostris* (Poey)', in *Lateral Line Detectors*, Cahn, P., ed., Indiana University Press, Bloomington, pp. 265–273.

Bardach, J. E., Todd, J. H. and Crickmer, R. (1967) Orientation by taste in fish of the genus *Ictalurus*. *Science*, **155**, 1276–1278.

Best, A. C. G. and Nicol, J. A. C. (1980) Eyeshine in fishes. A review of ocular reflectors. *Can. J. Zool.*, **58**, 945–956.

Blaxter, J. H. S., Denton, E. J. and Gray, J. A. B. (1981) 'Acousticolateralis system in clupeid fishes', in *Hearing and Sound Communication in Fishes*, Tavolga, W. N., Popper, A. N. and Fay, R. R., eds., Springer-Verlag, New York, pp. 39–59.

Bone, Q. and Ryan, K. P. (1978) Cupular sense organ in *Ciona* (Tunica: Ascidiascea). *J. Zool.* (Lond.), **186**, 417–427.

Buerkle, U. (1968) Relation of pure tone threshold to background noise level in the Atlantic cod (*Gadus morhua*). *J. Fish. Res. Bd. Can.*, **25**, 1115–1160.

Denton, E. J. (1970) On the organization of reflecting surfaces in some marine animals. *Phil. Trans. Roy. Soc. Lond. B*, **178**, 285–313.

Denton, E. J., Gilpin-Brown, J. B. and Wright, P. G. (1972) The angular distribution of light produced by some mesopelagic fish in relation to their camouflage. *Proc. Roy. Soc. Lond. B*, **182**, 145–158.

Denton, E. J. and Gray, J. A. B. (1979) The analysis of sound by the sprat ear. *Nature*, **282**, 406–407.

Enger, P. S. (1967) Hearing in herring. *Comp. Biochem. Physiol.*, **22**, 527–538.

Fänge, R. (1982) Exogenous otoliths of elasmobranchs. *J. mar. biol. Ass. U.K.*, **62**, 225.

Fletcher, A., Murphy, T. and Young, A. (1954) Solution of two optical problems. *Proc. Roy. Soc. Lond. A*, **223**, 216.

Gruber, S. H. (1975) Duplex vision in the elasmobranchs: histological, electrophysiological and psychophysical evidence, in *Vision in Fishes. New Approaches in Research*, Ali, M., ed., Plenum Press, New York, pp. 525–540.

Harris, G. G. and van Bergeijk, W. A. (1962) Evidence that the lateral-line organ responds to near-field displacements of sound sources in water. *J. Acoust. Soc. Amer.*, **34**, 1831–1841.

Herring, P. J. (1977) Bioluminescence of marine organisms. *Nature*, **267**, 788–793.

Herring, P. J. and Morin, J. G. (1978). 'Bioluminescence in fish, in *Bioluminescence in Action*, Herring, P. J., ed., Academic Press, London, pp. 273–329.

Hudspeth, A. J. and Jacobs, R. (1979) Stereocilia mediate transduction in vertebrate hair cells. *Proc. Natl. Acad. Sci. U.S.A.*, **76**, 1506–1509.

Iversen, R. T. B. (1967) 'Response of yellowfin tuna (*Thunnus albacares*) to underwater sounds', in *Proc. 2nd Symp. Marine Bio-acoustics*, Vol. 2, Pergamon Press, Oxford, pp. 105–121.

Kalmijn, A. J. (1978) 'Electric and magnetic sensory world of sharks, skates and rays', in *Sensory Biology of Sharks, Skates and Rays*, Hodgson, E. S. and Mathewson, R. F., eds., U.S. Dept. of the Navy, office of Naval Research, Arlington, Va., pp. 507–528.

Lawry, J. V. (1974) Lantern fish compare downwelling light and bioluminescence. *Nature*, **247**, 155–157.

Liebman, P. A. (1972) 'Microspectrophotometry of photoreceptors', in *Handbook of Sensory Physiology*, Vol. 7, pt. 1, *Photochemistry of Vision*, Dartnall, H. J. A., ed., Springer-Verlag, Berlin, pp. 481–528.

Lissmann, H. W. (1958) On the function and evolution of electric organs in fish. *J. exp. Biol.*, **35**, 156–191.

Lissmann, H. W. (1963) Electrolocation by fishes. *Sci. Amer.*, **209**, 50–59.

Lowenstein, O. and Wersäll, J. (1959) A functional interpretation of the electron microscopic structure of the sensory brain in the cristae of the elasmobranch *Raja clavata* in terms of directional sensitivity. *Nature*, **184**, 1807–1808.

Lythgoe, J. N. (1979) *The Ecology of Vision*. Clarendon Press, Oxford.

McFarland, W. N. and Muntz, F. W. (1975) The evolution of photopic visual pigments in fishes. *Vision Res.*, **15**, 1071–1080.

Muir-Evans, H. (1940) *Brain and Body of Fish*. Technical Press, London.

Muntz, W. R. A. (1976) On yellow lenses in mesopelagic animals. *J. mar. biol. Ass. U.K.*, **56**, 963–976.

Nicol, J. A. C., Arnott, H. J. and Best, A. C. G. (1973) Tapeta lucida in bony fishes (Actinopterygii): a survey. *Can. J. Zool.*, **51**, 69–81.

Pickens, P. E. and McFarland, W. N. (1964) Electric discharge and associated behaviour in the stargazer. *Animal Behaviour*, **12**, 362–367.

Pumphrey, R. J. (1950) 'Hearing', in *Symp. soc. exp. Biol.*, no. 4, Cambridge University Press, pp. 3–18.

Pumphrey, R. J. (1961) 'Concerning vision', in *The Cell and the Organism*, Ramsey, J. A. and Wigglesworth, V. B., eds., Cambridge University Press, pp. 193–208.

Ridge, R. M. A. P. (1977) Physiological responses of stretch receptors in the pectoral fin of the ray *Raja clavata*. *J. mar. biol. Ass. U.K.*, **57**, 535–541.

Somiya, H. (1976) Functional significance of the yellow lens in the eyes of *Argyropelecus affinis*. *Mar. biol.*, **34**, 93–99.

Somiya, H. and Tamura, T. (1973) Studies on the visual accommodation in fishes. *Jap. J. Ichthyol.*, **20**, 193–206.

Szabo, T., Kalmijn, A. J., Enger, P. S. and Bullock, T. H. (1972) Microampullary organs and a submandibular sense organ in the freshwater ray, *Potamotrygon*. *J. comp. Physiol.*, **79**, 15–27.

von Frisch, K. (1938) Uber das Gehörsinn der Fische. *Biol. Revs.*, **11**, 210–246.

von Frisch, K. (1938) Zur Psychologie des Fisch-Schwarmes. *Naturwiss.*, **26**, 601–606.

Walls, G. L. (1942) *The Vertebrate Eye and its Adaptive Radiation*. Cranbrook Inst. Sci., Michigan.

Chapter 10

Aronson, L. R. (1981) 'Evolution of telencephalic function in lower vertebrates', in *Brain Mechanisms of Behaviour in Lower Vertebrates*, Laming, P. R., ed., Cambridge University Press, pp. 33–58.

Bullock, T. H. (1982) Electroreception. *Ann Rev. Neurosci.*, **5**, 121–170.

Cohen, A. H. and Wallén, P. (1980) The neuronal correlate of locomotion in fish. 'Fictive swimming' induced in an *in vitro* preparation of the lamprey spinal cord. *Exp. Brain Res.*, **41**, 11–18.

Davis, R. E., Kassel, J. and Martinez, M. (1981) 'The telencephalon and reproductive behaviour in the teleost *Macropodus opercularis* (L.): effects of lesions on the incidence of spawning and egg cannibalism', in *Brain Mechanisms of Behaviour in Lower Vertebrates*, Laming, P. R., ed., Cambridge University Press, pp. 239–255.

De Bruin, J. P. C. (1980) 'Telencephalon and behaviour in teleost fish. A neuroethological approach', in *Comparative Neurology of the Telencephalon*, Ebbesson, S. O., ed., Plenum Press, New York.

Demski, L. S. (1981) 'Hypothalamic mechanisms of feeding in fishes', in *Brain Mechanisms of Behaviour in Lower Vertebrates*, Laming, P. R., ed., Cambridge University Press, pp. 225–237.

Eaton, R. C. and Bombardieri, R. A. (1978) 'Behavioural functions of the Mauthner neuron', in *Neurobiology of the Mauthner Cell*, Faber, D. and Korn, H., eds., Raven Press, New York, pp. 221–244.

Ebbesson, S. O. E. and Northcutt, R. G. (1976) 'Neurology of anamniotic vertebrates', in *Evolution of Brain and Behaviour in Vertebrates*, Masterton, R. B., Bullerman, M. E., Campbell, C. B. G. and Hotton, N., eds., Lawrence Erlbaum Assoc. Inc., Hillsdale, N. J., pp. 115–146.

Eccles, J. C. (1973) *The understanding of the brain*. McGraw-Hill, New York.

Echteler, S. M. and Saidel, W. M. (1981) Forebrain connections in the goldfish support telencephalic homologies with land vertebrates. *Science*, 212, 683–685.

Finger, T. E., Bell, C. C. and Russell, C. J. (1981) Electrosensory pathways to the valvulae cerebelli in mormyrid fish. *Exp. Brain Res.*, 42, 23–33.

Flood, N. B. and Overmier, J. B. (1981) 'Learning in teleost fish: role of the telencephalon', in *Brain Mechanisms of Behaviour in Lower Vertebrates*, Laming, P. R., ed., Cambridge University Press, pp. 259–279.

Gleibman, H. and Rozin, P. (1971) 'Learning and memory', in *Fish Physiology*, Vol. 6, Hoar, W. S. and Randall, D. J., eds., Academic Press, New York, pp. 191–278.

Graeber, R. C. (1978) 'Behavioural studies correlated with central nervous system integration of vision in sharks', in *Sensory Biology of Sharks, Skates and Rays*, Hodgson, E. S. and Mathewson, T. F., eds., U.S. Dept. of the Navy, Office of Naval Research, Arlington, Va., pp. 195–225.

Guthrie, D. M. (1981) 'The properties of the visual pathway of a common freshwater fish (*Perca fluviatilis*) in relation to visual behaviour', in *Brain Mechanisms of Behaviour in Lower Vertebrates*, Laming, P. R., ed., Cambridge University Press, pp. 79–112.

Guthrie, D. M. and Banks, J. R. (1978) The receptive field structure of visual cells from the optic tectum of the freshwater perch (*Perca fluviatilis*). *Brain Res.*, 141, 211–225.

Hultborn, H., Lindstrom, H. and Wigstrom, H. (1979) On the function of recurrent inhibition in the spinal cord. *Exp. Brain Res.*, 37, 399–403.

Ito, H., Tanaka, H., Sakamoto, N. and Morita, Y. (1981) Isthmic afferent neurons identified by the retrograde HRP method in a teleost, *Navodon modestus*. *Brain Res.*, 207, 163–169.

Kapoor, B. G., Evans, H. E. and Pevzner, R. A. (1975) The gustatory system in fish. *Adv. mar. Biol.*, 13, 53–108.

Kleerekoper, H. (1969) *Olfaction in Fishes*. Indiana University Press, Bloomington.

Marshall, N. B. (1971) *Explorations in the Life of Fishes*. Harvard University Press, Cambridge, Mass.

Moller, P. and Szabo, T. (1981) Lesions in the nucleus mesencephali exterolateralis: effects on electrocommunication in the mormyrid fish *Gnathonemus petersii* (Mormyriformes). *J. comp. Physiol.*, 144, 327–333.

Nieuwenhuys, R. (1964) Organisation of the spinal cord. *Progr. Brain Res.*, 11, 1–55.

Nieuwenhuys, R. (1967) Comparative anatomy of the cerebellum. *Progr. Brain Res.*, 25, 1–93.

Nieuwenhuys, R. and Nicholson, C. (1969) 'A survey of the general morphology, the fibre connections and the possible functional significance of the gigantocerebellum of mormyrid fishes', in *Neurobiology of Cerebellar Evolution and Development*, Llinas, R., ed., A.M.A., Chicago, pp. 107–134.

Nieuwenhuys, R., Pouwels, E. and Smulders-Kersten, E. (1974) The neuronal organization of cerebellar lobe C1 in the mormyrid fish *Gnathonemus Petersii* (Teleostei). *Z. Anat. Entw. Gesch.*, 144, 315–336.

Northcutt, R. G. (1978) 'Brain organization in the cartilaginous fishes', in *Sensory Biology of Sharks, Skates and Rays*, Hodgson, E. S. and Mathewson, R. F., eds., Office of Naval Research, Dept. of the Navy, Arlington, Va., pp. 117–193.

Northcutt, R. G. (1980) Anatomical evidence of electroreception in the coelacanth (*Latimeria chalumnae*). *Zbl. Vet. Med. C. Anat. Histol. Embryol.*, **9**, 289–295.

Northcutt, R. G. (1981) Evolution of the telencephalon in mammals. *Ann. Rev. Neurosci.*, **4**, 301–350.

Paul, D. H. and Roberts, B. L. (1979) The significance of cerebellar function for a reflex movement of the dogfish. *J. comp. Physiol.*, **134**, 69–74.

Poon, M. L. T. (1980) Induction of swimming in lamprey by L-DOPA and amino acids. *J. comp. Physiol.*, **136**, 337–344.

Roberts, B. L. (1981*a*) 'The organisation of the nervous system of fishes in relation to locomotion', in *Vertebrate Locomotion*, Day, M. H., ed., *Symp. zool. soc. Lond.*, no. 48, pp. 115–136.

Roberts, B. L. (1981*b*) Central processing of acousticolateralis signals in elasmobranchs, in *Hearing and Sound Communication in Fishes*, Tavolga, W. N., Popper, A. N. and Fay, R. R., eds., Springer-Verlag, New York, pp. 357–372.

Roberts, A. and Clerk, J. D. W. (1982) The neuroanatomy of an amphibian spinal cord. *Phil. Trans. Roy. Soc. B.*, **296**, 195–212.

Roberts, B. L. and Witkovsky, P. (1975) A functional analysis of the mesencephalic nucleus of the fifth nerve in the selachian brain. *Proc. Roy. Soc. Lond. B*, **190**, 473–495.

Rovainen, C. M. (1979) Neurobiology of lampreys. *Physiol. Revs.*, **59**, 1007–1077.

Savage, G. E. (1980) 'The fish telencephalon and its relation to learning', in *Comparative Neurology of the Telencephalon*, Ebbesson, S. O. E., ed., Plenum Press, New York, pp. 129–174.

Smith, C. U. M. (1971) *The Brain*. Faber and Faber, London.

Szarski, H. (1980) 'A functional and evolutionary interpretation of brain size in vertebrates', in *Evolutionary Biology*, Vol. 13, Hecht, M. K., Steere, W. C. and Wallace, B., eds., pp. 149–174.

Vanegas, H. (1981) 'The teleostean optic tectum: neuronal substrates for behavioural mechanisms', in *Brain Mechanisms of Behaviour in Lower Vertebrates*, Laming, P. R., ed., Cambridge University Press, pp. 113–121.

Whiting, H. P. W. (1948) Nervous structure of the spinal cord of the young larval brook-lamprey. *Quart. J. micr. Sci.*, **89**, 359–383.

Zipser, B. and Bennett, M. V. L. (1976) Interaction of electrosensory and electromotor signals in lateral line lobe of a mormyrid fish. *J. Neurophysiol.*, **39**, 713–721.

Zottoli, S. J. (1978) 'Comparative morphology of the Mauthner cell in fish and amphibians', in *Neurobiology of the Mauthner Cell*, Faber, D. S. and Korn, H., eds., Raven Press, New York, pp. 13–45.

Zottoli, S. J. (1981) Electrophysiological and morphological characterization of the winter flounder Mauthner cell. *J. comp. Physiol.*, **144**, 541–553.

Index